MEGA 015 Agriculture

Laurel K. Hoover

This page is intentionally left blank.

Agricultural Learning

Printed in the United States of America

This page is intentionally left blank.

Table of Content

This page is intentionally left blank.

Chapter 1 – Questions

QUESTION 1

Demonstrate knowledge of the history, rationale, and methods of agriscience education.Which historical event significantly influenced the development of agriscience education in the United States?

A. The invention of the cotton gin in 1793
B. The establishment of the Land Grant Act in 1862
C. The introduction of hybrid corn in the 1920s
D. The formation of the FFA (Future Farmers of America) in 1928

Answer:

QUESTION 2

Demonstrate knowledge of global issues affecting agriculture.Which factor poses the greatest threat to global food security in the coming decades?

A. Technological advancements in agriculture
B. Climate change and extreme weather events
C. Expansion of genetically modified crops
D. Increased international trade agreements

Answer:

QUESTION 3

Apply knowledge of animal nutrition and feeding practices.Which nutrient is essential for preventing white muscle disease in ruminants?

A. Calcium
B. Phosphorus
C. Vitamin D
D. Selenium

Answer:

QUESTION 4

In agriscience education, what safety practice is essential to prevent the spread of diseases among livestock?

A. Integrated Pest Management (IPM)
B. Hazard Communication
C. Biosecurity Measures
D. Lockout/Tagout Procedures

Answer:

QUESTION 5

Apply knowledge of skills related to career planning, job search, and job acquisition. During a job interview for an agriculture-related position, what is the most effective way to demonstrate adaptability?

A. Highlighting expertise in a single specialized area
B. Emphasizing the preference for routine and stability
C. Discussing experiences of successfully handling unexpected challenges
D. Avoiding any mention of previous job changes

Answer:

QUESTION 6

Identify types and characteristics of crop production practices.Which planting method is known for its ability to conserve water, reduce erosion, and promote soil health by leaving crop residues on the field?

A. No-till planting
B. Broadcast seeding
C. Double-row planting
D. Flood irrigation

Answer:

QUESTION 7

Demonstrate knowledge of types and characteristics of forests in Ohio and their uses.In Ohio's forests, what is the primary ecological function of the Eastern Hemlock (Tsuga canadensis)?

A. Soil erosion control
B. Nitrogen fixation
C. Carbon sequestration
D. Shade tolerance for understory plants

Answer:

QUESTION 8

Agriculture teacher Sarah is teaching her students about the integration of business concepts in crop production. A group of students is considering adopting a new irrigation system for their vegetable farm. To make a sound decision, which business concept should they prioritize?

A. Break-Even Analysis
B. Maslow's Hierarchy of Needs
C. Herzberg's Two-Factor Theory
D. Hertz's Rental Car Pricing Model

Answer:

QUESTION 9

A group of agriculture students is conducting an experiment to analyze factors affecting plant growth. They notice that plants in one group are exhibiting stunted growth despite receiving adequate water, light, and nutrients. What additional factor should they investigate to understand this anomaly?*

A. Soil Composition
B. Atmospheric Pressure
C. Genetic Factors
D. Moon Phases

Answer:

QUESTION 10

A group of agriculture students is troubleshooting a malfunction in a tractor's hydraulic system. The hydraulic lift is not operating properly. What could be a potential cause of this issue?

A. Low Engine Oil Level
B. Air in the Hydraulic Fluid
C. Overcharged Battery
D. Faulty Ignition Switch

Answer:

QUESTION 11

In agribusiness recordkeeping, what is the purpose of a depreciation schedule for assets?

A. Estimating future market values for asset resale.
B. Calculating tax deductions for business expenses.
C. Tracking the historical cost of assets over time.
D. Assessing the overall financial health of the business.

Answer:

QUESTION 12

Agriculture teacher Maria is tasked with developing a marketing plan for the school's agricultural products. What crucial step should Maria prioritize in the early stages of creating the marketing plan?

A. Launching the marketing campaign to build immediate awareness.
B. Identifying and understanding the target market for the agricultural products.
C. Developing test markets to assess the viability of the products.
D. Allocating the entire budget to promotional activities.

Answer:

QUESTION 13

A group of agriculture students is learning about safe animal handling practices. During a practical session, they need to move a herd of cattle from one pasture to another. What is the most effective way to ensure safe and stress-free movement?

A. Use loud noises to encourage faster movement.
B. Approach the herd quickly to establish dominance.
C. Move calmly and quietly, using low-stress handling techniques.
D. Use prods or sticks to guide individual animals.

Answer:

QUESTION 14

A group of agriculture teachers is discussing the use of cover crops in horticulture. Which characteristic makes a plant well-suited for use as a cover crop?

A. Slow Growth Rate
B. Shallow Root System
C. Nitrogen-Fixing Ability
D. High Water Requirements

Answer:

QUESTION 15

Weed, Disease, and Insect Pest Control - Case Study Situation:

Agriculture teacher Emma is teaching her students about integrated pest management in horticulture. In a case study, a farmer is dealing with an infestation of aphids on tomato plants. What is the most effective and environmentally friendly method for controlling aphids in this situation?

A. Broadcasting chemical insecticides.
B. Releasing ladybugs (predatory beetles) to feed on aphids.
C. Applying synthetic fertilizers to enhance plant resistance.
D. Pruning and removing affected plants.

Answer:

QUESTION 16

Agriculture teacher Olivia is discussing the environmental impact of animal production systems. In a case study, a large-scale poultry farm is facing criticism for its contribution to water pollution. What practice could be implemented to minimize nutrient runoff and protect water quality?

A. Implementing a comprehensive wastewater treatment system.
B. Increasing the density of poultry in existing facilities.
C. Fertilizing nearby fields with excess poultry litter.
D. Diverting wastewater directly into nearby water bodies.

Answer:

QUESTION 17

Agriculture teacher Mark is teaching students about the safe handling of confinement animal facilities. In a case study, a student is tasked with entering a swine confinement building. What precautionary measure is essential for the student's safety?

A. Entering the building alone for efficiency.
B. Wearing a mask only if a noticeable odor is present.
C. Providing proper ventilation before entering.
D. Ignoring biosecurity protocols for a quick visit.

Answer:

QUESTION 18

Importance and Scope of the Agriculture Industry. Which of the following agricultural practices contributes the LEAST to sustainable food production?

A. Aquaponics
B. No-till farming
C. Subsistence farming
D. Agroforestry

Answer:

QUESTION 19

Basic Ecological Principles and their Application to Agriculture. Which ecological principle involves the role an organism plays within its environment and includes its interactions with other organisms and resources?

A. Habitat partitioning
B. Trophic cascades
C. Ecological succession
D. Ecological niche

Answer:

QUESTION 20

Formation and Characteristics of Soil, Soil Structure, and Fertility. Which soil texture is generally considered most conducive to plant growth due to its balanced water retention and drainage characteristics?

A. Clay
B. Sand
C. Silt
D. Loam

Answer:

QUESTION 21

Apply knowledge of strategies for self-assessment, self-improvement, career exploration, and college and career readiness.What is a key advantage of engaging in informational interviews as part of the career exploration process?

A. It guarantees a job offer from the interviewee.
B. It allows for networking and building professional relationships.
C. It provides a detailed critique of your resume and cover letter.
D. It ensures immediate placement in the desired career.

Answer:

QUESTION 22

Tools, Materials, and Skills in Agricultural Structures. Which of the following materials provides the highest thermal insulation and is commonly used in constructing energy-efficient agricultural structures?

A. Concrete blocks
B. Metal panels
C. Straw bales
D. PVC pipes

Answer:

QUESTION 23

Soil Management, Conservation, and Erosion Control. Among the following practices, which is considered an effective method for controlling soil erosion on steep slopes?

A. Contour plowing
B. Broadcast seeding
C. Deep plowing
D. Overgrazing

Answer:

QUESTION 24

Apply knowledge of entrepreneurship and strategies for starting and managing an agribusiness.What is a potential challenge faced by agribusiness entrepreneurs who adopt a direct-to-consumer sales model?

A. Limited control over pricing.
B. Difficulty in accessing wholesale markets.
C. Lower profit margins compared to retail distribution.
D. Increased reliance on intermediaries.

Answer:

QUESTION 25

Demonstrate knowledge of underlying growth processes and stages of growth in animals.In the context of animal growth, what is a potential consequence of rapid growth during the early stages of an animal's life?

A. Improved feed efficiency.
B. Reduced risk of skeletal disorders.
C. Increased susceptibility to metabolic disorders.
D. Enhanced reproductive performance.

Answer:

QUESTION 26

Demonstrate knowledge of alternative energy systems in agriculture.In the context of wind energy for agricultural applications, what is a potential limitation of relying solely on wind power?

A. High installation costs.
B. Unpredictable wind patterns.
C. Limited adaptability to remote farm locations.
D. Minimal impact on reducing dependence on fossil fuels.

Answer:

QUESTION 27

Which plant propagation technique involves encouraging the development of roots on a stem that is still attached to the parent plant, followed by detachment to create a new individual?

A. Grafting
B. Layering
C. Division
D. Tissue culture

Answer:

QUESTION 28

Demonstrate knowledge of consumer behavior and the selling process.What is a potential challenge faced by agricultural sales professionals when dealing with consumers who engage in extensive online research before making a purchase?

A. Difficulty in establishing personal connections.
B. Limited access to product information.
C. Decreased price sensitivity.
D. Reduced reliance on customer reviews.

Answer:

QUESTION 29

Demonstrate knowledge of types, characteristics, and uses of renewable and nonrenewable natural resources, and principles and methods for their conservation and sustainable management.In the context of sustainable agriculture, what is a potential drawback of relying solely on monoculture cropping systems?

A. Enhanced biodiversity.
B. Reduced vulnerability to pests.
C. Improved soil structure.
D. Increased risk of pest outbreaks and disease epidemics.

Answer:

QUESTION 30

How does soil structure affect crop production?

A. It has no significant impact on crop production.
B. Good soil structure promotes root penetration and water infiltration.
C. Soil structure is only relevant for ornamental plants, not crops.
D. Poor soil structure enhances nutrient availability to crops.

Answer:

QUESTION 31

What is a key consideration in water management for greenhouse and nursery crops to prevent root diseases?

A. Frequent deep watering
B. Watering during the hottest part of the day
C. Overhead irrigation
D. Well-draining growing media

Answer:

QUESTION 32

What is a key consideration in the care of floral design tools and equipment to ensure their longevity and effectiveness?

A. Exposure to direct sunlight
B. Regular cleaning and sanitizing
C. Storing tools in a humid environment
D. Using tools for multiple purposes

Answer:

QUESTION 33

Which government agency is primarily responsible for addressing environmental issues and regulations related to agriculture in the United States?

A. Department of Agriculture (USDA)
B. Environmental Protection Agency (EPA)
C. Food and Drug Administration (FDA)
D. National Oceanic and Atmospheric Administration (NOAA)

Answer:

QUESTION 34

What drafting technique is crucial for ensuring accurate measurements and layout in shop fabrication?

A. Freehand sketching
B. Isometric projection
C. Orthographic projection
D. Parallel projection

Answer:

QUESTION 35

In the context of seedbed preparation, what is the primary purpose of utilizing a raised bed planting layout?

A. Enhanced water retention
B. Improved drainage
C. Increased soil compaction
D. Efficient weed control

Answer:

QUESTION 36

Which stage of growth and development in livestock is characterized by rapid muscle and bone development, making proper nutrition critical for long-term health?

A. Neonatal stage
B. Weaning stage
C. Growing stage
D. Mature stage

Answer:

QUESTION 37

What factor can impact customer relations in the sale of agricultural products?

A. Strict adherence to fixed pricing
B. Consistent product quality
C. Limited communication with customers
D. Flexible and responsive customer service

Answer:

QUESTION 38

A student is tasked with drafting detailed plans for a metalworking project involving the fabrication of a complex structure. The student is unsure whether to use isometric projection or orthographic projection for the drafting.Which projection method is more appropriate for conveying a three-dimensional understanding of the structure?

A. Isometric projection
B. Orthographic projection
C. Both methods can be used interchangeably.
D. It depends on the viewer's perspective

Answer:

When considering tillage practices, which method is aimed at minimizing soil disturbance and maintaining crop residues on the soil surface?

A. No-till
B. Moldboard plowing
C. Disk harrowing
D. Subsoiling

Answer:

QUESTION 40

What is a potential consequence of environmental degradation on agricultural production?

A. Increased productivity
B. Expansion of farmland
C. Enhanced aquifer replenishment
D. Decreased productivity

Answer:

QUESTION 41

When using oxyacetylene cutting in metalworking, what is a key consideration to prevent the oxidation of the cut edge?

A. Increased oxygen flow
B. Slower cutting speed
C. Lower acetylene pressure
D. Post-cut cooling with water

Answer:

QUESTION 42

What is the primary function of sulfur in plant nutrition?

A. Enhances root development
B. Stimulates flower and fruit formation
C. Aids in photosynthesis
D. Improves water retention in cells

Answer:

QUESTION 43

When constructing a landscape feature in California, which material choice aligns with sustainable practices by minimizing environmental impact?

A. Pressure-treated wood
B. Concrete
C. Recycled composite materials
D. Vinyl fencing

Answer:

QUESTION 44

In livestock digestion, where does the majority of microbial fermentation take place?

A. Stomach
B. Small intestine
C. Cecum
D. Large intestine

Answer:

QUESTION 45

In soil erosion control, which vegetative practice is particularly effective in reducing water runoff and promoting water infiltration?

A. Monoculture planting
B. Contour plowing
C. Overgrazing
D. Clear-cutting

Answer:

QUESTION 46

When designing an agricultural structure, what is a critical consideration for siting to ensure optimal functionality and efficiency?

A. Proximity to water sources
B. Accessibility to main roads
C. Aesthetics and visual appeal
D. Elevation above sea level

Answer:

QUESTION 47

Among the agriculturally important crops grown in California, which one is a legume crop primarily used for forage and soil improvement?

A. Wheat
B. Alfalfa
C. Corn
D. Rice

Answer:

QUESTION 48

In animal breeding, what term refers to the phenomenon where the combination of two different alleles at a specific gene locus results in a phenotype that is superior to the phenotypes of either homozygous individual?

A. Dominance
B. Epistasis
C. Heterosis
D. Pleiotropy

Answer:

QUESTION 49

In entrepreneurship, what term describes the ability to identify and capitalize on new and innovative opportunities in the agricultural sector?

A. Risk aversion
B. Adaptability
C. Resource allocation
D. Opportunity recognition

Answer:

QUESTION 50

You are tasked with designing an agricultural structure on a hilly terrain. The land is prone to erosion, and there are concerns about soil stability. Which construction method would be most effective in minimizing erosion and ensuring the structural integrity of the building?

A. Constructing a building foundation using a slab-on-grade method.
B. Implementing contour plowing around the structure to redirect water flow.
C. Utilizing timber framing for the structure to enhance aesthetic appeal.
D. Installing a French drain system around the perimeter of the building.

Answer:

QUESTION 51

Among the common California crops, which harvesting method requires the most delicate handling to avoid damaging the product during collection?

A. Wheat (combine harvesting)
B. Oranges (handpicking)
C. Lettuce (mechanical cutting)
D. Grapes (mechanical harvesting)

Answer:

QUESTION 52

When selecting a landscape plant for an area prone to drought conditions, which characteristic should be given the highest priority?

A. Rapid growth rate
B. High water requirement
C. Deep taproot system
D. Large foliage surface area

Answer:

QUESTION 53

Among the following, which type of animal facility often utilizes a slatted floor system to manage waste and maintain cleanliness?

A. Dairy barn
B. Poultry house
C. Swine farrowing house
D. Cattle feedlot

Answer:

QUESTION 54

Among the various sources of credit available to agricultural businesses, which one typically requires collateral and is often used for long-term investments?

A. Trade credit
B. Operating loan
C. Intermediate-term loan
D. Real estate mortgage loan

Answer:

QUESTION 55

What is the primary purpose of disbudding or dehorning cattle in livestock management?

A. Enhancing their aesthetic appearance
B. Preventing damage to fencing and facilities
C. Reducing the risk of injuries to other animals
D. Improving the animals' social behavior

Answer:

QUESTION 56

What is a critical aspect to consider when conducting prescribed burns in forest and wildland ecosystems?

A. Complete suppression of fire
B. Burning during high humidity
C. Burning only during the night
D. Adherence to predetermined weather conditions

Answer:

QUESTION 57

Among the types of internal combustion engines used in agricultural machinery, which engine cycle is commonly found in smaller, handheld power equipment such as chainsaws or string trimmers?

A. Four-stroke cycle
B. Diesel cycle
C. Rotary engine cycle
D. Two-stroke cycle

Answer:

QUESTION 58

Which environmental factor significantly impacts the thermal comfort and well-being of poultry, especially in hot climates?

A. High humidity levels
B. Increased ventilation
C. Direct sunlight exposure
D. Low stocking density

Answer:

QUESTION 59

In agricultural business accounting, what is the primary purpose of conducting a cash flow analysis?

A. To determine profitability
B. To assess liquidity
C. To evaluate asset values
D. To calculate depreciation

Answer:

QUESTION 60

A farmer intends to mow a large pasture area covered with tall grass and weeds. The primary objective is to efficiently cut and evenly distribute the cut material for proper decomposition without damaging the pasture surface. Which agricultural power equipment is most appropriate for this task?

A. Disc harrow
B. Flail mower
C. Grain combine
D. Rotary cutter

Answer:

QUESTION 61

A farmer is considering implementing precision farming practices on their large-scale crop production. Which technology would be most beneficial for the farmer to optimize resource use and increase overall efficiency?

A. Traditional planting methods
B. Randomized irrigation
C. Variable Rate Technology (VRT)
D. Manual harvesting

Answer:

QUESTION 62

During a lesson on meat inspection, agriculture teachers highlight the importance of post-mortem examination. What aspect of meat inspection is addressed through post-mortem examination?

A. External appearance
B. Internal temperature
C. Visible fat content
D. Disease detection

Answer:

QUESTION 63

What is the primary function of transpiration in plants?

A. Nutrient absorption
B. Water transport
C. Photosynthesis
D. Hormone production

Answer:

QUESTION 64

In a floral design class, students are learning about the role of filler flowers in arrangements. Which flower would be most suitable as a filler due to its small and delicate appearance?

A. Sunflowers
B. Baby's breath (Gypsophila)
C. Orchids
D. Peonies

Answer:

QUESTION 65

A group of students is studying the interrelationship between climate and plant communities. In a region with a Mediterranean climate, which adaptation would you expect in the plant community?

A. Drought-resistant plants with deep root systems
B. Shade-loving plants with large leaves
C. Cold-tolerant plants with short growing seasons
D. Water-loving plants with shallow root systems

Answer:

QUESTION 66

Agricultural researchers are exploring biotechnological approaches for crop improvement. If a new genetically engineered crop is developed to resist a specific pest, what is a potential concern that agriculture teachers should address when discussing this technology with students?

A. Increased yield
B. Biodiversity loss
C. Improved taste
D. Soil fertility

Answer:

QUESTION 67

Agriculture teachers are discussing methods for preventing and treating external pests in livestock. Which practice is effective in reducing the risk of external parasite infestations in cattle?

A. Providing shaded areas
B. Regular hoof trimming
C. Rotational grazing
D. Increasing protein in the diet

Answer:

QUESTION 68

Agriculture teachers are discussing the impact of government policies on agricultural businesses. If a government imposes a tariff on imported agricultural products, what is the likely effect on domestic farmers?

A. Increased competition
B. Decreased production
C. Higher prices for domestic products
D. Expanded export opportunities

Answer:

QUESTION 69

In an agricultural mechanics class, students are learning about the safe use of power tools. What precaution is most important when using a chainsaw to cut wood?

A. Wearing sturdy gloves
B. Wearing eye protection
C. Ensuring proper ventilation
D. Using noise-canceling headphones

Answer:

QUESTION 70

In an agricultural workshop, students are learning about the safe use of tractors. A farmer is preparing to plow a field, and there is a slight slope. What precaution should the farmer take to ensure tractor safety on the slope?

A. Increase speed for quicker plowing
B. Use the tractor with worn-out brakes
C. Drive straight up and down the slope
D. Carry heavy loads on the uphill side

Answer:

QUESTION 71

In an equipment maintenance class, students are learning about the importance of regular inspections. A farmer notices that the safety shields on a hay baler are missing. What action should the farmer take to address this safety concern?

A. Continue operating the baler without the safety shields.
B. Fabricate makeshift shields from available materials.
C. Replace the missing safety shields with the manufacturer-approved parts.
D. Ignore the issue, as safety shields are not essential for baler operation.

Answer:

QUESTION 72

Demonstrate knowledge of types and characteristics of forests in Ohio and their uses.Which tree species, commonly found in Ohio forests, is classified as a hardwood and is known for its high-quality wood used in furniture production?

A. Eastern White Pine (Pinus strobus)
B. Black Walnut (Juglans nigra)
C. Red Oak (Quercus rubra)
D. Sugar Maple (Acer saccharum)

Answer:

QUESTION 73

A group of agriculture teachers is participating in a horticultural workshop. During a plant identification exercise, they come across a plant with lobed leaves, milky sap, and a hollow stem. What is the correct classification of this plant?

A. Herbaceous Annual
B. Biennial Shrub
C. Perennial Tree
D. Herbaceous Perennial

Answer:

QUESTION 74

Agriculture teacher Jessica is planning a field trip to a local farm with her students. The farm produces a variety of agricultural products, including fruits, vegetables, and livestock. As she develops the educational program for the field trip, what strategy should Jessica employ to effectively showcase the diversity of agricultural products to her students?

A. Focus exclusively on the farm's highest revenue-generating products.
B. Design interactive stations highlighting different aspects of crop and livestock production.
C. Skip the farm visit and opt for classroom discussions on agricultural diversity.
D. Emphasize only the most popular agricultural products in the region.

Answer:

QUESTION 75

Career Opportunities in the Agriculture Industry. Among the following career roles, which one typically involves precision technology and data analysis in modern agriculture?

A. Agricultural inspector
B. Crop consultant
C. Livestock breeder
D. Agronomist

Answer:

QUESTION 76

Tools, Materials, and Skills in Agricultural Structures. Among the following tools, which is commonly used for the maintenance of agricultural structures to test electrical conductivity in soils and ensure proper grounding?

A. Multimeter
B. Plumb bob
C. Hacksaw
D. Torque wrench

Answer:

QUESTION 77

Demonstrate knowledge of underlying growth processes and stages of growth in animals.In the growth process of animals, what is a potential drawback of relying solely on high-energy diets to accelerate weight gain?

A. Improved muscle development.
B. Enhanced feed efficiency.
C. Increased risk of metabolic disorders.
D. Decreased feed conversion rates.

Answer:

QUESTION 78

In greenhouse and nursery management, what is the purpose of using shade cloth?

A. To reduce water consumption
B. To control temperature
C. To prevent pest infestations
D. To increase humidity

Answer:

QUESTION 79

Demonstrate knowledge of principles and methods for forest management.When implementing a shelterwood harvesting method in forest management, what is the primary goal of leaving some mature trees in place?

A. Enhancing wildlife habitat
B. Promoting natural regeneration
C. Maximizing timber production
D. Minimizing the risk of forest fires

Answer:

QUESTION 80

Case Study - Transplanting Techniques

A horticulture class is tasked with transplanting seedlings into larger containers. The students are debating whether it's necessary to remove the seedlings from their original containers before transplanting. What is the most appropriate advice?

A. No, transplanting can be done without removing the seedlings.
B. Yes, always remove seedlings from their original containers before transplanting.
C. It depends on the type of seedling and its root structure.
D. Only remove seedlings if they are root-bound.

Answer:

QUESTION 81

Agriculture teacher Ryan is leading a student-led project to market locally grown produce from the school's greenhouse. The team is developing a marketing plan, and they are discussing market segmentation. What approach should Ryan recommend for effective market segmentation in this context?

A. Targeting all community members with a general marketing message.
B. Dividing the market based on age, income, and lifestyle to tailor marketing efforts.
C. Ignoring market segmentation and marketing the produce universally.
D. Focusing solely on advertising to schools and educational institutions.

Answer:

QUESTION 82

Which contemporary issue in agriculture is characterized by the use of technology and data-driven solutions to optimize farming practices and increase efficiency?

A. Sustainable Agriculture
B. Precision Farming
C. Urban Agriculture
D. Agroecology

Answer:

QUESTION 83

Agriculture teachers are discussing the principles of animal housing with their students. One teacher emphasizes providing adequate ventilation in livestock housing. Why is proper ventilation crucial for animal health?

A. To prevent escape attempts by animals.
B. To control the spread of infectious diseases.
C. To reduce noise levels within the facility.
D. To discourage aggressive behavior among animals.

Answer:

QUESTION 84

Agriculture teacher Alex is discussing how plants absorb nutrients from the soil. In a case study, students learn about a crop suffering from stunted growth despite adequate nutrient availability in the soil. What factor could be impeding nutrient uptake in this situation?

A. Low soil pH.
B. Excessive soil moisture.
C. High soil organic matter content.
D. Presence of mycorrhizal fungi.

Answer:

QUESTION 85

Formation and Characteristics of Soil, Soil Structure, and Fertility. Which soil horizon is most likely to contain the highest amount of organic matter and humus?

A. A horizon (topsoil)
B. B horizon (subsoil)
C. C horizon (parent material)
D. R horizon (bedrock)

Answer:

QUESTION 86

Demonstrate knowledge of characteristics and purposes of different types of business organizations.In the context of agriculture, which characteristic best distinguishes a cooperative from a partnership?

A. Limited liability for members.
B. Equal decision-making power for all members.
C. Ability to raise capital through the sale of stock.
D. Profit-sharing based on individual contributions.

Answer:

QUESTION 87

When considering the use of biomass for energy in agriculture, what is a potential concern related to sustainability?

A. Reduced dependence on fossil fuels.
B. Increased demand for land and competition with food crops.
C. Enhanced soil fertility through biomass decomposition.
D. Mitigation of greenhouse gas emissions.

Answer:

QUESTION 88

Which floral design element refers to the visual and tactile quality of the surface of the materials used in an arrangement?

A. Line
B. Texture
C. Form
D. Space

Answer:

QUESTION 89

When working with local advisory committees in agriscience education, what strategy is focused on gathering input and insights from various stakeholders to make informed decisions for program improvement?

A. Top-Down Approach
B. Needs Assessment
C. Experiential Learning
D. Direct Instruction

Answer:

QUESTION 90

Agriculture teacher Mark is discussing nutrient deficiencies in plants with his students. In a case study, a farmer observes yellowing of the leaves in a crop. What nutrient deficiency is most likely causing this symptom?

A. Nitrogen deficiency.
B. Phosphorus deficiency.
C. Potassium deficiency.
D. Iron deficiency.

Answer:

QUESTION 91

Development of Agricultural Products, Value-added Principles, and Distribution. Which practice directly contributes to enhancing the value-added aspect of agricultural products?

A. Reducing the use of organic fertilizers
B. Implementing efficient packaging methods
C. Opting for conventional farming techniques
D. Minimizing product diversification

Answer:

QUESTION 92

Apply knowledge of strategies for self-assessment, self-improvement, career exploration, and college and career readiness.What is a potential drawback of relying solely on aptitude tests for self-assessment in the context of agriculture careers?

A. Aptitude tests are time-consuming.
B. Aptitude tests may not consider personal interests and values.
C. Aptitude tests are too subjective.
D. Aptitude tests are only applicable to specific agricultural fields.

Answer:

QUESTION 93

Demonstrate knowledge of alternative energy systems in agriculture.In the context of alternative energy sources for agriculture, what is a potential drawback of relying solely on solar power for farm operations?

A. Inconsistent energy production due to weather conditions.
B. High upfront installation costs.
C. Limited adaptability to different farm sizes.
D. Minimal impact on reducing greenhouse gas emissions.

Answer:

QUESTION 94

When selecting flowers for a floral arrangement, what characteristic is important for ensuring a visually appealing design?

A. Uniform color
B. Same stem length
C. Homogeneous texture
D. Varied heights and shapes

Answer:

QUESTION 95

A group of students is developing a new value-added food product for a local farmers' market. They are debating whether to use a cost-plus pricing strategy or a penetration pricing strategy. Which consideration is essential for making this decision?

A. Brand Loyalty
B. Elasticity of Demand
C. SWOT Analysis
D. Pareto Principle

Answer:

QUESTION 96

How does effective communication contribute to building a positive learning environment in agriscience education?

A. Using jargon to establish authority and command respect.
B. Providing vague instructions to encourage student independence.
C. Encouraging open and honest dialogue to address concerns.
D. Avoiding nonverbal cues to maintain a professional image.

Answer:

QUESTION 97

Agriculture teacher Rachel is discussing ways to minimize environmental damage in animal production. In a case study, a hog farm is dealing with odor issues affecting the surrounding community. What practice could be employed to address this concern?

A. Increasing the concentration of hog pens in close proximity.
B. Installing air filtration systems in residential areas.
C. Implementing proper waste management and odor control measures.
D. Disposing of hog waste in nearby water bodies.

Answer:

QUESTION 98

Apply knowledge of characteristics and uses of different types of fertilizers and other components of crop growth systems.Agricultural teachers often emphasize the importance of balanced fertilizer application. What is a potential drawback of relying solely on organic fertilizers for crop nutrition?

A. Risk of soil compaction.
B. Uneven nutrient distribution.
C. Slow release of nutrients.
D. High susceptibility to leaching.

Answer:

QUESTION 99

Which plant propagation method involves the removal of a portion of a stem with one or more buds attached and its subsequent planting to generate a new plant?

A. Grafting
B. Layering
C. Budding
D. Cutting

Answer:

QUESTION 100

What is a potential environmental impact of monoculture in agricultural production?

A. Increased biodiversity
B. Enhanced soil fertility
C. Runoff of fertilizers and pesticides
D. Improved water table recharge

Answer:

QUESTION 101

Which crop management practice is a form of cultural control and involves the strategic arrangement of different crops in a specific sequence over several seasons to minimize pests and diseases?

A. Mulching
B. Crop rotation
C. Irrigation
D. Land classification

Answer:

QUESTION 102

When managing a forest ecosystem for recreation purposes, what principle emphasizes the careful planning of trails and recreational facilities to minimize negative impacts on the natural environment?

A. Sustainable logging
B. Leave No Trace
C. Clear-cutting
D. Habitat fragmentation

Answer:

QUESTION 103

Which cellular structure is responsible for regulating the movement of substances in and out of plant cells, controlling cell turgor pressure?

A. Tonoplast
B. Plasmodesmata
C. Cell wall
D. Plasma membrane

Answer:

QUESTION 104

In terms of waste management, which practice is commonly employed to effectively reduce odor and pathogens in stored manure?

A. Aerating the manure
B. Adding more bedding material
C. Covering the manure storage area
D. Increasing the storage temperature

Answer:

QUESTION 105

A farmer needs to harvest a large area of mature wheat while minimizing grain loss and ensuring a clean, efficient harvest. The crop is mature and stands tall. Which agricultural power equipment would be most efficient for this harvest?

A. Hay baler
B. Plow
C. Grain combine
D. Seed drill

Answer:

QUESTION 106

In a swine production class, farmers are learning about common internal pests affecting pigs. If a pig is displaying symptoms such as coughing and labored breathing, what internal parasite should be considered as a potential cause?

A. Tapeworms
B. Liver flukes
C. Lungworms
D. Roundworms

Answer:

QUESTION 107

What is a key principle in landscape design that involves arranging plantings and hardscapes to create visually appealing and balanced outdoor spaces?

A. Unity
B. Scale
C. Emphasis
D. Proportion

Answer:

QUESTION 108

When developing a business plan for an agricultural enterprise, what section typically outlines the company's goals, mission, and strategies for achieving a competitive advantage?

A. Financial projections
B. Marketing plan
C. Executive summary
D. Operations plan

Answer:

QUESTION 109

What is a primary concern regarding the post-harvest handling of leafy greens like spinach or lettuce?

A. Ethylene exposure
B. Temperature fluctuations
C. Moisture content
D. Pathogen contamination

Answer:

QUESTION 110

Which tool or equipment is primarily used in timber cruising for estimating the volume and value of standing timber?

A. GPS receiver
B. Compass
C. Increment borer
D. Tree caliper

Answer:

QUESTION 111

Which environmental factor has the most significant impact on the rate of plant respiration?

A. Light intensity
B. Temperature
C. Soil moisture
D. Atmospheric pressure

Answer:

QUESTION 112

Agriculture teachers are discussing sustainable resource use. What principle should be emphasized when explaining the concept of sustainable agriculture to students?

A. Maximizing resource extraction for short-term gain
B. Minimizing biodiversity to enhance crop yields
C. Balancing resource use to meet present needs without compromising future generations
D. Focusing solely on the economic profitability of agricultural practices

Answer:

QUESTION 113

Among the following, which is an example of biological pest control in agriculture?

A. Synthetic pesticides
B. Crop rotation
C. Ladybugs for aphid control
D. Herbicide application

Answer:

QUESTION 114

Which wildlife management strategy involves manipulating the environment to create or enhance habitat features to benefit specific wildlife species?

A. Exotic species introduction
B. Predator control
C. Habitat manipulation
D. Population control

Answer:

QUESTION 115

Which plant organ is primarily responsible for the storage of carbohydrates and other essential nutrients in most vascular plants?

A. Roots
B. Stems
C. Leaves
D. Flowers

Answer:

QUESTION 116

What primary function does a biosecurity protocol serve in animal production systems?

A. Minimizing waste generation
B. Optimizing feed efficiency
C. Reducing disease transmission risks
D. Enhancing animal reproduction rates

Answer:

QUESTION

Which of the following is NOT a factor influencing plant growth?

A. Pruning
B. Soil pH
C. Atmospheric pressure
D. Phototropism

Answer:

QUESTION 117

Agriculture teachers are discussing the concept of niche in an ecosystem. Which scenario best illustrates the concept of a niche?

A. A forest where various tree species coexist.
B. A river with diverse fish species.
C. A grassland where different herbivores graze.
D. A desert with specialized cactus species.

Answer:

QUESTION 118

In landscape planning, what term refers to the process of selecting and placing plants based on their mature size, form, and compatibility to achieve a balanced and well-proportioned design?

A. Xeriscaping
B. Plant zoning
C. Biomimicry
D. Planting design

Answer:

QUESTION 119

In business management functions, which activity involves determining the organizational structure, allocating resources, and defining responsibilities?

A. Organizing
B. Planning
C. Directing
D. Controlling

Answer:

QUESTION120

Which food safety issue associated with crop production primarily pertains to concerns about the potential adverse effects of chemical residues on consumers?

A. Irradiated food
B. Genetic modification of crops
C. Organic farming practices
D. Pesticide residues

Answer:

QUESTION 121

What primary function does a clinometer serve in forest management?

A. Measuring tree diameter
B. Assessing soil moisture levels
C. Estimating tree height or slope steepness
D. Determining timber species

Answer:

QUESTION 122

In a scenario where a farmer is considering adopting micropropagation techniques for propagating a rare and endangered plant species, which factor should the farmer carefully evaluate before making a decision?

A. Soil composition
B. Genetic diversity
C. Atmospheric pressure
D. Water availability

Answer:

QUESTION 123

In an agricultural economics class, students are discussing the concept of elasticity. If the demand for a specific crop is inelastic, what does this imply about consumers?

A. Consumers are very responsive to changes in price.
B. Consumers are not sensitive to changes in price.
C. Consumers are indifferent to the quantity available.
D. Consumers prefer substitute products.

Answer:

QUESTION 124

In the context of the reproductive system in livestock, what is the primary function of the corpus luteum?

A. Ovulation
B. Sperm production
C. Milk production
D. Fetal development

Answer:

QUESTION 125

When working with electrical systems in agriculture, what is the primary function of a ground fault circuit interrupter (GFCI)?

A. Regulating voltage
B. Preventing electrical shock
C. Controlling current flow
D. Distributing power evenly

Answer:

QUESTION 126

Among the following, which equipment is commonly used for installing sod or turfgrass in landscaping projects?

A. Dethatcher
B. Lawn roller
C. Aerator
D. Sod cutter

Answer:

QUESTION 127

In production management, what critical factor does market forecasting primarily aid agricultural businesses in?

A. Estimating labor costs
B. Predicting equipment depreciation
C. Planning production quantities
D. Calculating soil fertility levels

Answer:

QUESTION 128

A floriculture class is arranging floral designs for an exhibition. The instructor emphasizes the importance of using appropriate flower varieties for line, mass, filler, and form. What type of flower is most suitable for creating a sense of movement and direction in a floral arrangement?

A. Roses
B. Tulips
C. Lilies
D. Snapdragons

Answer:

QUESTION 129

In the context of animal breeding, what is the primary advantage of utilizing artificial insemination (AI)?

A. Increased genetic diversity
B. Enhanced reproductive efficiency
C. Promotion of natural selection
D. Prevention of genetic mutations

Answer:

QUESTION 130

You are tasked with planning an irrigation system for a large agricultural field with diverse crops. The goal is to optimize water use efficiency and minimize water wastage. What irrigation method would be most suitable for achieving these objectives?

A. Flood irrigation with furrows.
B. Overhead sprinkler irrigation.
C. Drip irrigation with precision emitters.
D. Subsurface drip irrigation.

Answer:

QUESTION 131

What is the primary purpose of using anthelmintics in livestock management?

A. To control external parasites
B. To treat bacterial infections
C. To prevent viral diseases
D. To control internal parasites (worms)

Answer:

QUESTION 132

Which agricultural power equipment is commonly equipped with a Power Take-Off (PTO) shaft used for transferring power to implements such as balers, mowers, or augers?

A. Plow
B. Disc harrow
C. Grain combine
D. Rotary cutter

Answer:

QUESTION 133

A livestock farmer notices a sudden decrease in egg production in the poultry flock. Upon inspection, there are no apparent signs of illness or distress. What could be a potential hidden factor affecting egg production?

A. Lack of ventilation
B. Presence of internal parasites
C. Insufficient lighting
D. Low ambient temperature

Answer:

QUESTION 134

Agriculture teacher Sarah is teaching her students about plant genetics. She presents a case study where a farmer notices variations in the height and color of crops within the same field, despite planting identical seeds. What genetic phenomenon is likely responsible for this variability?

A. Inbreeding depression.
B. Hybrid vigor.
C. Genetic mutation.
D. Gene silencing.

Answer:

QUESTION 135

Basic Principles of Product Pricing and Promotion in Agriculture. Which pricing strategy involves setting an initial high price for a new agricultural product and gradually reducing it over time?

A. Penetration pricing
B. Price skimming
C. Psychological pricing
D. Bundle pricing

Answer:

QUESTION 136

Which of the following resources is most likely to provide insights into the day-to-day challenges faced by individuals in specific agriculture careers?

A. A glossy promotional brochure from a seed company.
B. A peer-reviewed research article on sustainable farming practices.
C. An autobiography of a successful farmer.
D. An advertisement for agricultural machinery.

Answer:

QUESTION 137

Demonstrate knowledge of proper management of agricultural waste products and practices used to protect soil, air, and water quality.When implementing conservation tillage practices, what is a potential benefit for soil health and water conservation?

A. Increased soil erosion.
B. Enhanced water runoff.
C. Improved water infiltration and retention.
D. Decreased organic matter content in the soil.

Answer:

QUESTION 138

In floral design, what principle refers to the visual equality of elements and their arrangement, creating a sense of stability and harmony?

A. Contrast
B. Scale
C. Balance
D. Proportion

Answer:

QUESTION 139

When selecting a welding process for a specific metalworking task, what factor is essential to consider in order to achieve a strong and durable weld?

A. Welding speed
B. Ambient temperature
C. Welding rod color
D. Metal compatibility

Answer:

QUESTION 140

Agriculture teacher Michael is discussing methods for handling and processing harvested horticultural products. In a case study, students learn about a fruit orchard experiencing post-harvest losses due to improper handling. What practice should be implemented to minimize post-harvest losses?

A. Delaying transportation to allow fruits to ripen further.
B. Sorting and grading fruits based on quality and ripeness.
C. Storing fruits in open-air containers.
D. Applying wax coatings on fruits for extended shelf life.

Answer:

QUESTION 141

Demonstrate knowledge of characteristics and purposes of different types of business organizations.Which aspect is a unique characteristic of a limited liability company (LLC) compared to a sole proprietorship in the context of agribusiness?

A. Limited access to business loans.
B. Personal liability for business debts.
C. Pass-through taxation.
D. Single owner control.

Answer:

QUESTION 142

Which of the following soil components is primarily responsible for retaining water and providing nutrients to plants?

A. Sand
B. Clay
C. Silt
D. Humus

Answer:

QUESTION 143

Which factor has the most significant impact on the nutritional requirements of livestock, considering factors such as species, age, pregnancy, and lactation?

A. Environmental temperature
B. Geographic location
C. Feed storage method
D. Optimal weight gain

Answer:

QUESTION 144

In a metalworking workshop, a student is tasked with welding two pieces of metal. The metal pieces are identified as aluminum and carbon steel. The student is considering using the same welding process for both.Which action should the student take to ensure successful welding?

A. Proceed with the same welding process for both aluminum and carbon steel.
B. Adjust the welding process parameters for each metal type.
C. Choose a different welding process for aluminum.
D. Weld aluminum first and then carbon steel.

Answer:

QUESTION 145

Identify types and characteristics of crop production practices.Which crop rotation strategy is designed to break pest and disease cycles by alternating crops with different growth requirements?

A. Monoculture
B. Polyculture
C. Sequential cropping
D. Diversified cropping

Answer:

QUESTION 146

Case Study - Agricultural Knowledge in Business Decision Making

A group of agriculture students is involved in a project where they need to decide between conventional and organic farming practices for a specific crop. Which agricultural knowledge should be considered in this decision-making process?

A. Time Value of Money
B. Plant Physiology
C. Human Resource Management
D. Six Sigma Methodology

Answer:

QUESTION 147

What is a crucial element of leadership in agriscience education when managing a team of teachers?

A. Micromanaging each teacher's tasks to ensure consistency.
B. Allowing teachers complete autonomy without guidance.
C. Providing constructive feedback and support for professional development.
D. Ignoring individual teacher strengths and weaknesses.

Answer:

QUESTION 148

Agriculture teacher Sarah is instructing her students on the safe operation of a tractor. In a case study, a student is attempting to start the tractor but encounters difficulty. What is the safest course of action for the student?

A. Bypassing the ignition system to force the tractor to start.
B. Continuously attempting to start the tractor until it engages.
C. Investigating and troubleshooting the issue before attempting to start again.
D. Asking a nearby friend to assist without checking the tractor's condition.

Answer:

QUESTION 149

Effects of Agriculture on the Environment and Production Systems. Among the following production systems, which is most likely to result in increased soil erosion and nutrient runoff?

A. Conventional agriculture
B. Organic agriculture
C. Sustainable agriculture
D. Permaculture

Answer:

QUESTION 150

Demonstrate knowledge of types of weeds, diseases, and insect pests affecting crops and methods for their control.When managing a crop with a high susceptibility to a particular insect pest, which integrated pest management (IPM) strategy is most effective in preventing resistance development?

A. Rotating insecticides with similar modes of action.
B. Using a single, potent insecticide consistently.
C. Introducing natural predators to control the pest population.
D. Applying insecticides only after pest thresholds are exceeded.

Answer:

QUESTION 151

Apply knowledge of skills related to career planning, job search, and job acquisition.When negotiating salary for an agriculture-related position, what is the most appropriate approach?

A. Immediately stating a desired salary to set expectations
B. Waiting for the employer to propose a salary range first
C. Focusing solely on the base salary and ignoring benefits
D. Accepting the first offer without negotiation

Answer:

QUESTION 152

Which type of Supervised Agricultural Experience (SAE) is characterized by owning and operating an agriculture-related business, often involving the production or sale of goods or services?

A. Exploratory SAE
B. Placement SAE
C. Entrepreneurship SAE
D. Research SAE

Answer:

QUESTION 153

A group of agriculture teachers is planning a unit on animal health. They want to emphasize preventive measures for common diseases. One teacher suggests using vaccines as a primary preventive tool. Another teacher argues that maintaining optimal environmental conditions is equally important. What concept combines both approaches?

A. Adaptive Immunity
B. Herd Immunity
C. Biosecurity
D. Antimicrobial Resistance

Answer:

QUESTION 154

Apply knowledge of sources of information about agriculture careers.Which of the following sources is likely to provide the most accurate and up-to-date information about current trends and demands in the agriculture job market?

A. A textbook published five years ago.
B. An online agriculture forum with active participation from industry professionals.
C. A documentary film on traditional farming practices.
D. A pamphlet distributed by a local agricultural equipment supplier.

Answer:

QUESTION 155

Demonstrate knowledge of proper management of agricultural waste products and practices used to protect soil, air, and water quality.When considering the disposal of agricultural plastic waste, what is a potential environmental concern associated with burning these materials on the farm?

A. Reduced greenhouse gas emissions.
B. Release of toxic fumes and air pollutants.
C. Enhanced soil fertility.
D. Prevention of water pollution.

Answer:

QUESTION 156

In the context of the total agricultural program model, what is the primary purpose of supervised agricultural experiences (SAEs)?

A. To provide students with hands-on, real-world learning opportunities
B. To fulfill academic requirements without practical application
C. To encourage competition among students
D. To assess theoretical knowledge through exams

Answer:

QUESTION 157

Demonstrate knowledge of standard and alternative methods for animal production and management.In poultry production, what is the primary benefit of utilizing the deep-litter system?

A. Improved egg production
B. Reduced ammonia levels in the coop
C. Minimized risk of disease transmission
D. Enhanced growth rate of broilers

Answer:

QUESTION 158

What physiological process in plants is directly influenced by the availability of oxygen in the root zone?

A. Photosynthesis
B. Transpiration
C. Respiration
D. Transport of Nutrients

Answer:

QUESTION 159

Importance and Scope of the Agriculture Industry. Which of the following factors is NOT a primary contributor to the sustainability of modern agriculture?

A. Crop rotation
B. Monoculture practices
C. Integrated pest management
D. Soil conservation techniques

Answer:

QUESTION 160

Irrigation and Drainage Systems in Agricultural Production. Which irrigation system typically requires the highest initial investment but offers precise water delivery, minimizing water wastage in agricultural fields?

A. Flood irrigation
B. Drip irrigation
C. Sprinkler irrigation
D. Furrow irrigation

Answer:

QUESTION 161

Apply knowledge of principles of animal genetics and animal reproduction and their applications in agriculture.When implementing selective breeding for a specific trait in livestock, what is a challenge associated with relying solely on closely related animals for mating?

A. Increased genetic diversity.
B. Improved resistance to diseases.
C. Higher likelihood of genetic disorders.
D. Accelerated growth rates.

Answer:

QUESTION 162

Demonstrate knowledge of plant classification and characteristics and uses of various species of plants.Among the following plant species, which one is both a fruit and a vegetable, causing confusion in its classification?

A. Tomato (Solanum lycopersicum)
B. Apple (Malus domestica)
C. Banana (Musa spp.)
D. Strawberry (Fragaria × ananassa)

Answer:

QUESTION 163

A local farmer is considering diversifying their crop production to meet the growing demand for organic produce. After conducting a market analysis, they discover a high demand for organic tomatoes. However, they are unsure about the potential risks and benefits. What economic principle is crucial for the farmer to consider in this decision-making process?

A. Law of Diminishing Marginal Utility
B. Law of Supply and Demand
C. Law of Comparative Advantage
D. Law of Increasing Opportunity Cost

Answer:

QUESTION 164

Agriculture teachers are conducting a workshop on equipment maintenance. One teacher suggests regular lubrication as a key practice for preventing machinery breakdowns. Another teacher argues that it's more crucial to focus on proper calibration of electronic components. What is the most balanced approach?

A. Prioritize regular lubrication over electronic component calibration.
B. Emphasize electronic component calibration without regular lubrication.
C. Both regular lubrication and electronic component calibration are equally important.
D. Focus on neither regular lubrication nor electronic component calibration.

Answer:

QUESTION 165

What is the primary advantage of using a soilless growing medium in greenhouse and nursery production?

A. Enhanced nutrient content
B. Reduced water retention
C. Lower cost
D. Increased disease resistance

Answer:

QUESTION 166

What is a potential challenge in marketing agricultural products internationally due to cultural factors?

A. Homogeneous consumer preferences worldwide
B. Standardized packaging and labeling
C. Variations in dietary preferences and habits
D. Consistent marketing strategies

Answer:

QUESTION 167

When managing outdoor recreation areas, what is a key consideration to balance the needs of visitors and the preservation of the natural environment?

A. Implementing strict access restrictions
B. Encouraging unregulated exploration
C. Promoting intensive development
D. Adopting sustainable management practices

Answer:

QUESTION 168

Which of the following agricultural chemical application methods poses the highest risk of off-target drift and potential environmental contamination?

A. Granular application
B. Foliar spray
C. Seed treatment
D. Soil injection

Answer:

QUESTION 169

In forest management, which social factor significantly influences the adoption of sustainable practices and conservation efforts?

A. Ethnic diversity
B. Economic incentives
C. Educational background
D. Urbanization rates

Answer:

QUESTION 170

In the process of photosynthesis, which organelle plays a central role?

A. Vacuole
B. Chloroplast
C. Mitochondrion
D. Endoplasmic reticulum

Answer:

QUESTION 171

A shop fabrication project involves cutting intricate designs into a sheet of metal. The student has the options of using either plasma cutting or oxyacetylene cutting for this task.Which cutting method is more suitable for achieving precise and intricate designs in the metal?

A. Plasma cutting
B. Oxyacetylene cutting
C. Both methods are equally suitable.
D. It depends on the thickness of the metal sheet.

Answer:

QUESTION 172

When evaluating a livestock carcass, what factor is NOT typically considered in assessing meat quality?

A. Marbling
B. Muscling
C. Feathering
D. Yield grade

Answer:

QUESTION 173

In carpentry, what is the purpose of a "dado" joint?

A. Joining two pieces of wood at an angle
B. Creating a strong joint between two perpendicular pieces
C. Producing a joint with increased surface area for gluing
D. Joining two pieces of wood end-to-end

Answer:

QUESTION 174

Which tool is specifically designed for the accurate trimming and shaping of shrubs and hedges in landscape maintenance?

A. Hand trowel
B. Pruning shears
C. Hedge trimmers
D. Lawn edger

Answer:

QUESTION 175

In leadership styles, what approach emphasizes collaboration, communication, and team involvement, seeking input from all team members in decision-making processes?

A. Autocratic leadership
B. Laissez-faire leadership
C. Transformational leadership
D. Democratic leadership

Answer:

QUESTION 176

Which of the following irrigation methods is generally considered the most water-efficient for landscape plants?

A. Sprinkler irrigation
B. Drip irrigation
C. Flood irrigation
D. Subsurface irrigation

Answer:

QUESTION 177

Demonstrate knowledge of student organizations, including the role of faculty advisors.What is the primary responsibility of a faculty advisor for a student agricultural organization?

A. Facilitating social events for students
B. Providing academic tutoring
C. Guiding students in setting and achieving organizational goals
D. Managing the organization's finances

Answer:

QUESTION 178

When considering emerging technologies in agricultural production, which technology is designed to optimize irrigation efficiency?

A. Unmanned aerial vehicles (UAVs) for crop monitoring.
B. Precision agriculture sensors for soil health assessment.
C. Drones equipped with thermal imaging for pest detection.
D. Automated drip irrigation systems with real-time data monitoring.

Answer:

QUESTION 179

Development of Agricultural Products, Value-added Principles, and Distribution. Among the following, which process is crucial in the value-added chain of agricultural products?

A. Cross-breeding livestock for desirable traits
B. Incorporating artificial preservatives into food products
C. Utilizing renewable energy sources in farming practices
D. Implementing blockchain technology for supply chain traceability

Answer:

QUESTION 180

A group of agriculture teachers is discussing the integration of various systems in modern agricultural machinery. A teacher argues that a combine harvester primarily utilizes an internal combustion engine. Another teacher disagrees, stating that electrical systems are equally essential for its operation. What is the correct perspective on this matter?

A. Combines rely exclusively on internal combustion engines.
B. Combines primarily use electrical systems for harvesting.
C. Combines integrate both internal combustion and electrical systems.
D. Combines are solely powered by hydraulic systems.

Answer:

QUESTION 181

Demonstrate knowledge of structure and function of cells, tissues, and plant processes.Which organelle is responsible for the synthesis of ATP (adenosine triphosphate) in plant cells?

A. Chloroplast
B. Vacuole
C. Nucleus
D. Mitochondrion

Answer:

QUESTION 182

A group of farmers is considering adopting a new technology for pest control on their crops. The technology is relatively untested in the local area. How should the farmers approach the risk analysis of this decision?

A. Focus on Historical Cost Analysis
B. Conduct a Sensitivity Analysis
C. Ignore Potential Risks for Innovation
D. Rely on Rule-Based Decision Making

Answer:

QUESTION 183

Characteristics and Uses of Various Animal Species. Which of the following animal species is considered a "hindgut fermenter"?

A. Cattle
B. Horses
C. Dogs
D. Rabbits

Answer:

QUESTION 184

Demonstrate knowledge of consumer behavior and the selling process.Which sales strategy is most effective in addressing the needs of agricultural consumers who prioritize sustainability and environmental responsibility?

A. Highlighting cost savings and discounts.
B. Emphasizing product features and specifications.
C. Showcasing the product's eco-friendly certifications and sustainable practices.
D. Offering exclusive promotions and limited-time offers.

Answer:

QUESTION 185

In terms of soil texture, which type is known for having the smallest particle size and high fertility due to its ability to retain nutrients?

A. Sandy soil
B. Loamy soil
C. Clayey soil
D. Silt soil

Answer:

QUESTION 186

Demonstrate knowledge of standard and alternative methods for animal production and management.In animal husbandry, what is the primary purpose of utilizing rotational grazing practices?

A. Increasing overall pasture productivity
B. Minimizing the need for veterinary care
C. Promoting natural selection among the herd
D. Eliminating the use of supplemental feed

Answer:

QUESTION 187

Which environmental factor has a direct impact on both transpiration and photosynthesis in plants?

A. Soil pH
B. Wind Speed
C. Soil Moisture Content
D. Soil Nutrient Levels

Answer:

QUESTION 188

In agribusiness accounting, what is the primary purpose of a cash flow statement?

A. Assessing the company's overall profitability.
B. Tracking the movement of money in and out of the business.
C. Evaluating the market value of agricultural assets.
D. Predicting long-term investment returns.

Answer:

QUESTION 189

In the context of leadership in agriscience education, what is a characteristic of a transformational leader?

A. Strict adherence to established procedures and protocols.
B. Maintaining a hierarchical and authoritative approach.
C. Inspiring and motivating others to exceed their own expectations.
D. Avoiding involvement in hands-on activities with students.

Answer:

QUESTION 190

Sustainable Animal Production - Case Study Situation:

Agriculture teacher Daniel is discussing strategies for sustainable animal production. In a case study, a cattle ranch is seeking methods to reduce greenhouse gas emissions. What practice could contribute to achieving this goal?

A. Introducing methane-producing bacteria in cattle feed.
B. Implementing rotational grazing practices.
C. Increasing the use of synthetic growth hormones.
D. Encouraging sedentary behavior in cattle.

Answer:

QUESTION 191

Anatomy and Physiology of Animals. Among the following, which organ in avian anatomy functions similarly to the human liver?

A. Crop
B. Gizzard
C. Cloaca
D. Proventriculus

Answer:

QUESTION 192

When considering soil food chains, which organism plays a crucial role in decomposing organic matter and contributing to the formation of humus?

A. Earthworms
B. Bacteria
C. Fungi
D. Nematodes

Answer:

QUESTION 193

Demonstrate knowledge of characteristics and uses of different types of fertilizers and other components of crop growth systems.In a crop management scenario, what is a potential consequence of over-relying on nitrogen-based fertilizers?

A. Reduced soil acidity.
B. Increased susceptibility to disease.
C. Enhanced root development.
D. Altered nutrient balance in the soil.

Answer:

QUESTION 194

What is a key advantage of sexual reproduction in plants compared to asexual reproduction?

A. Greater genetic diversity
B. Faster propagation
C. Cloning of parent plants
D. Consistent traits in offspring

Answer:

QUESTION 195

Apply knowledge of skills related to career planning, job search, and job acquisition.When preparing a resume for a student seeking a career in agriculture, what is the most effective strategy for presenting technical skills?

A. Use industry-specific jargon to showcase expertise
B. Create a separate section exclusively for technical skills
C. Minimize technical details to focus on general qualifications
D. Provide lengthy explanations for each technical skill

Answer:

QUESTION 196

Apply knowledge of information technology and applications in agribusiness management.In the context of agribusiness management, what is the primary purpose of blockchain technology?

A. Enhancing crop yield through precision agriculture
B. Streamlining supply chain transparency and traceability
C. Facilitating communication between farmers and consumers
D. Optimizing irrigation systems for water conservation

Answer:

QUESTION 197

Agriculture teacher Jason is discussing greenhouse management with his students. In a case study, a nursery is struggling with excessive humidity, leading to fungal diseases in plants. What management practice should be implemented to address this issue?

A. Increasing the frequency of watering.
B. Installing additional heating systems.
C. Improving air circulation and ventilation.
D. Using chemical fungicides regularly.

Answer:

QUESTION 198

Agriculture teacher John is guiding his students in a simulated agribusiness project. The students are seeking financing for a new agricultural venture. They need to decide between a traditional bank loan and a venture capital investment. Which principle of finance is relevant to their decision-making process?

A. Time Value of Money
B. Net Present Value
C. Risk-Return Tradeoff
D. Cost of Capital

Answer:

QUESTION 199

Demonstrate knowledge of structure and function of cells, tissues, and plant processes.In plant cells, what is the primary function of the central vacuole?

A. Photosynthesis
B. Storage of nutrients and waste products
C. DNA replication
D. Cellular respiration

Answer:

QUESTION 200

What is a key aspect of effective communication in agriscience education?

A. Utilizing complex scientific terminology to demonstrate expertise.
B. Tailoring communication to the audience's level of understanding.
C. Avoiding visual aids to maintain a professional demeanor.
D. Minimizing feedback to maintain control of the conversation.

Answer:

QUESTION 201

Demonstrate knowledge of consumer behavior and the selling process.In the context of agricultural product sales, which psychological factor is most likely to influence a consumer's decision to purchase a product?

A. Perceived scarcity.
B. Product durability.
C. Packaging aesthetics.
D. Price consistency.

Answer:

QUESTION 202

Characteristics and Uses of Various Animal Species. Which of the following animal species has a gland known as the "brood patch" essential for incubating eggs?

A. Swine
B. Cattle
C. Poultry
D. Dogs

Answer:

QUESTION 203

Apply knowledge of information technology and applications in agribusiness management.What is the potential drawback of relying solely on automated data collection systems for farm management?

A. Increased efficiency in monitoring and analysis
B. Limited adaptability to changing environmental conditions
C. Enhanced precision in resource utilization
D. Decreased risk of data security breaches

Answer:

QUESTION 204

Apply knowledge of food safety issues and practices in crop production.To minimize the risk of contamination in crop production, what is the most effective method for preventing the spread of foodborne pathogens?

A. Using chemical pesticides liberally
B. Relying solely on organic farming practices
C. Implementing strict hygiene and sanitation measures
D. Increasing the use of genetically modified crops

Answer:

QUESTION 205

Agriculture teacher Ryan is discussing the characteristics of greenhouses. In a case study, a greenhouse is experiencing temperature fluctuations that impact plant growth. What structural modification could be implemented to regulate the temperature more effectively?

A. Installing additional overhead lighting for increased warmth.
B. Adding insulation to the greenhouse walls and roof.
C. Increasing the number of ventilation windows.
D. Reducing the overall size of the greenhouse.

Answer:

QUESTION 206

Basic Ecological Principles and their Application to Agriculture. What term describes the process of nutrient recycling in an ecosystem where nutrients are returned to the soil through the decomposition of organic matter?

A. Nitrogen fixation
B. Biomagnification
C. Denitrification
D. Nutrient cycling

Answer:

QUESTION 207

In selective breeding for crop improvement, what term refers to the process of crossing two genetically different individuals to produce offspring with desirable traits?

A. Hybridization
B. Self-pollination
C. Cloning
D. Mutation

Answer:

QUESTION 208

Which greenhouse and nursery crop is commonly categorized as a bedding plant?

A. Tomato
B. Poinsettia
C. Rose
D. Petunia

Answer:

QUESTION 209

In assessing the nutritional value of feed for livestock, what is the primary consideration related to the composition of the feed?

A. Color
B. Smell
C. Moisture content
D. Nutrient content

Answer:

QUESTION 210

What is a common symptom of nutrient deficiency in livestock that can be observed by agriculture teachers?

A. Increased appetite
B. Weight loss
C. Aggressive behavior
D. Shiny coat

Answer:

QUESTION 211

In sustainable agriculture, what is a key principle aimed at minimizing environmental degradation?

A. Crop rotation
B. Monoculture
C. Intensive pesticide use
D. Excessive irrigation

Answer:

QUESTION 212

In developing a marketing plan for agricultural products, what is a key consideration for identifying target markets?

A. Uniformity in consumer demographics
B. Ignoring competitors in the target market
C. Overlooking consumer trends
D. Tailoring products to meet specific consumer needs

Answer:

QUESTION 213

How do governmental factors affect international trade in agricultural products?

A. They have no impact on international trade.
B. They provide standardized regulations for all countries.
C. They introduce trade barriers and regulations.
D. They promote open-market policies globally.

Answer:

QUESTION 214

In metalworking, what property of wood makes it unsuitable for certain fabrication tasks?

A. Durability
B. Insulating properties
C. Conductivity
D. Combustibility

Answer:

QUESTION 215

Which of the following nutrients is primarily responsible for promoting flower and fruit development in plants?

A. Nitrogen
B. Phosphorus
C. Potassium
D. Calcium

Answer:

QUESTION 216

When conducting a soil test, which factor is NOT typically assessed to determine soil fertility?

A. pH level
B. Organic matter content
C. Electrical conductivity
D. Atmospheric pressure

Answer:

QUESTION 217

Which type of fertilizer is characterized by a gradual, sustained release of nutrients over an extended period, providing a slow and steady supply to plants?

A. Liquid fertilizer
B. Granular fertilizer
C. Inorganic fertilizer
D. Organic fertilizer

Answer:

QUESTION 218

When assessing livestock for breeding purposes, what trait is an indicator of the animal's ability to produce offspring with desirable characteristics?

A. Phenotype
B. Genotype
C. Heterosis
D. Heritability

Answer:

QUESTION 219

In plumbing, what is the purpose of a P-trap in a drainage system?

A. Regulating water pressure
B. Preventing backflow of sewage
C. Providing a smooth flow for water drainage
D. Removing impurities from water

Answer:

QUESTION 220

Among the following, which breed of cattle is primarily recognized for its superior milk production and is commonly associated with high butterfat content in its milk?

A. Angus
B. Holstein
C. Charolais
D. Limousin

Answer:

QUESTION 221

What is the primary purpose of using hedging in agricultural risk management?

A. Speculating on future price movements
B. Diversifying investment portfolios
C. Locking in a price for a future sale or purchase
D. Insuring against crop failure

Answer:

QUESTION 222

Which agricultural power equipment implements the use of a continuously variable transmission (CVT) system to optimize engine performance and fuel efficiency?

A. Rotary tiller
B. Tractor
C. Hay baler
D. Seed drill

Answer:

QUESTION 223

When selecting plants for a California landscape, which factor is critical to consider in order to conserve water and promote sustainable landscaping?

A. Rate of growth
B. Flower color
C. Drought tolerance
D. Soil acidity

Answer:

QUESTION 224

During meiosis, what is the result of independent assortment of chromosomes?

A. Formation of gametes with identical genetic material
B. Creation of somatic cells with varied genetic content
C. Segregation of homologous chromosomes into different gametes
D. Production of diploid cells with double the chromosome number

Answer:

QUESTION 225

You are overseeing the construction of an agricultural building that will house livestock. Proper ventilation is essential to maintain air quality and prevent respiratory issues in animals. Which design feature would contribute most effectively to ensuring adequate ventilation in the building?

A. Installing large, fixed windows on one side of the building.
B. Incorporating a ridge vent along the entire length of the roof.
C. Using airtight construction materials to enhance insulation.
D. Adding exhaust fans in the corners of the building.

Answer:

QUESTION 226

Which of the following symptoms is commonly associated with Foot-and-Mouth Disease (FMD) in livestock?

A. Hoof abscesses
B. Jaundice
C. Blister-like lesions on the mouth and feet
D. Respiratory distress

Answer:

QUESTION 227

What component is responsible for transmitting power from the engine to various machinery components in an agricultural power transmission system?

A. Hydraulic pump
B. Drive shaft
C. Clutch
D. Gearbox

Answer:

QUESTION 228

When caring for animals during gestation, what nutritional factor is particularly important to ensure proper fetal development?

A. High protein intake
B. Low carbohydrate diet
C. Adequate calcium supplementation
D. Proper energy balance

Answer:

QUESTION 229

Which of the following characteristics distinguishes monocots from dicots?

A. Presence of netted venation in leaves
B. Seeds with two cotyledons
C. Vascular bundles arranged in a ring
D. Parallel venation in leaves

Answer:

QUESTION 230

A farmer is preparing to cultivate a large field for planting crops. The terrain is slightly rugged and uneven, requiring equipment that can efficiently break up soil clods and prepare a smooth seedbed. Which agricultural power equipment would be most suitable for this task?

A. Hay baler
B. Rotary tiller
C. Seed drill
D. Plow

Answer:

QUESTION 231

During a lecture on private and public organizations in agriculture, students are exploring the role of agricultural cooperatives. What is a key advantage of farmers participating in agricultural cooperatives?

A. Increased individual risk
B. Limited access to markets
C. Enhanced bargaining power
D. Reduced need for government subsidies

Answer:

QUESTION 232

In a case study where a farmer is using Geographic Information Systems (GIS) and Global Positioning Systems (GPS) for crop management, what is a potential advantage that these technologies can offer?

A. Increased labor costs
B. Uniform application of pesticides
C. Limited field visibility
D. Decreased data accuracy

Answer:

QUESTION 233

During a lesson on securing and hauling loads, agriculture teachers are discussing the importance of weight distribution. What principle should farmers follow to ensure proper weight distribution when hauling loads on a trailer?

A. Placing the heaviest items in the center
B. Loading heavier items on one side for balance
C. Concentrating weight on the rear end
D. Distributing weight evenly across the trailer

Answer:

QUESTION 234

Agriculture teachers are teaching students about the proper storage of cut flowers. What aspect should they highlight as crucial to preventing bacterial growth and ensuring the longevity of cut flowers in storage?

A. Excessive sunlight exposure
B. High humidity levels
C. Warm storage temperatures
D. Clean water and sanitation

Answer:

QUESTION 235

Agriculture teachers are discussing the safe use of hand tools on the farm. What practice helps prevent accidents when using hand tools such as hammers and screwdrivers?

A. Using tools with worn-out handles
B. Wearing loose clothing
C. Keeping tools in a disorganized manner
D. Using the right tool for the task

Answer:

QUESTION 236

In a lesson on renewable energy sources, agriculture teachers are discussing the advantages and disadvantages of wind power. What environmental factor is most critical for the successful implementation of wind power?

A. Temperature variation
B. Wind speed and consistency
C. Soil composition
D. Solar radiation

Answer:

QUESTION 237

During a lesson on material disposal, agriculture teachers are discussing the proper disposal of chemical containers. A farmer has empty pesticide containers and is considering burning them to minimize waste. What advice should the teacher provide regarding the disposal of pesticide containers?

A. Burning the containers is an environmentally friendly disposal method.
B. Burying the containers in a designated area is a suitable option.
C. Recycling the containers through an approved program is recommended.
D. Throwing the containers in regular trash bins is acceptable.

Answer:

QUESTION 238

In a discussion on subsidies in agricultural business, agriculture teachers emphasize the importance of understanding the unintended consequences. If a government provides subsidies for a specific crop, what potential consequence should farmers be aware of?

A. Increased competition from imports
B. Overproduction and surplus
C. Decline in soil fertility
D. Reduced demand for the subsidized crop

Answer:

QUESTION 239

When discussing procedures for harvesting cut flowers, why is it important for agriculture teachers to emphasize the proper timing of harvest?

A. To reduce labor costs
B. To extend vase life
C. To promote root development
D. To enhance foliage color

Answer:

QUESTION 240

In a chemical safety workshop, agriculture teachers are emphasizing the safe use, storage, and disposal of materials. What precaution is essential when handling and storing fuels in an agricultural setting?

A. Storing fuels in a well-ventilated area
B. Using glass containers for fuel storage
C. Mixing different types of fuels for efficiency
D. Storing fuels near heat sources

Answer:

Chapter 2 – Answers and Explanations

QUESTION 1

Answer: B

Explanation: The Land Grant Act, also known as the Morrill Act, provided federal land to states to establish agricultural and mechanical colleges. This act laid the foundation for agricultural education in the U.S., leading to the development of agriscience education.

QUESTION 2

Answer: B

Explanation: Climate change and extreme weather events, such as droughts, floods, and storms, pose a significant threat to global food security. These events can disrupt agricultural production, affecting crop yields and food availability on a global scale.

QUESTION 3

Answer: D

Explanation: Selenium is crucial for preventing white muscle disease, a condition that affects the muscles and can lead to weakness and difficulty in movement in ruminants. Adequate selenium supplementation is necessary for proper muscle function.

QUESTION 4

Answer: C

Explanation: Biosecurity measures are protocols and practices implemented to prevent the introduction and spread of diseases. This is crucial in agriscience education to protect the health and well-being of livestock and maintain a safe learning environment.

QUESTION 5

Answer: C

Explanation: Demonstrating adaptability in an agriculture job interview involves sharing real-life examples of successfully navigating and overcoming unexpected challenges. This showcases the candidate's ability to adapt to varying circumstances in the field.

QUESTION 6

Answer: A

Explanation: No-till planting involves minimal soil disturbance, leaving crop residues on the field. This practice helps conserve water, reduce erosion, and enhance soil health by maintaining organic matter.

QUESTION 7

Answer: A

Explanation: Eastern Hemlock plays a crucial role in controlling soil erosion due to its dense and evergreen foliage. Its root systems stabilize soil, making it effective in preventing erosion in Ohio's forest ecosystems.

QUESTION 8

Answer: A

Explanation: In the context of crop production, Break-Even Analysis is crucial for determining the point at which the revenue from the vegetable sales equals the costs associated with the new irrigation system. This analysis helps in understanding the financial viability of the investment.

QUESTION 9

Answer: C

Explanation: Genetic factors play a significant role in plant growth. Even with optimal environmental conditions, different plant varieties may exhibit variations in growth patterns. Investigating genetic factors is crucial to understanding and addressing stunted growth in plants.

QUESTION 10

Answer: B

Explanation: Air in the hydraulic fluid can impede the proper functioning of the hydraulic system, affecting the performance of components like the lift. Bleeding the system to remove air is a common solution to restore functionality.

QUESTION 11

Answer: B

Explanation: A depreciation schedule in agribusiness recordkeeping is used for calculating tax deductions associated with the wear and tear of assets over time. This helps businesses accurately account for the decreasing value of their assets and manage tax liabilities.

QUESTION 12

Answer: B

Explanation: In the early stages of creating a marketing plan, Maria should prioritize identifying and understanding the target market for the agricultural products. This foundational step ensures that subsequent marketing efforts are tailored to the specific needs and preferences of the intended audience, increasing the plan's overall effectiveness.

QUESTION 13

Answer: C

Explanation: Low-stress handling techniques, such as calm and quiet movements, help reduce stress in animals during handling. This not only ensures their well-being but also promotes a safer and more controlled movement.

QUESTION 14

Answer: C

Explanation: A key benefit of using cover crops in horticulture is their ability to fix nitrogen in the soil, enhancing fertility. Plants with nitrogen-fixing abilities, such as legumes, are often chosen as cover crops to improve soil health and nutrient levels.

QUESTION 15

Answer: B

Explanation: Releasing ladybugs is a form of biological control that involves introducing natural predators to manage pest populations. Ladybugs are known to feed on aphids, providing an environmentally friendly solution to pest control in horticulture.

QUESTION 16

Answer: A

Explanation: Implementing a comprehensive wastewater treatment system is a crucial practice to minimize nutrient runoff from poultry farms. This helps mitigate environmental damage by treating and managing wastewater before it enters water bodies, protecting water quality.

QUESTION 17

Answer: C

Explanation: Providing proper ventilation before entering the swine confinement building is crucial for the student's safety. Adequate ventilation helps reduce the concentration of airborne contaminants, ensuring a safer environment for both animals and individuals entering the facility.

QUESTION 18

Answer: C

Explanation: While subsistence farming fulfills immediate food needs for a household, it often lacks advanced techniques for sustainable practices seen in aquaponics, no-till farming, and agroforestry. It may involve traditional methods that may not prioritize long-term sustainability or environmental conservation.

QUESTION 19

Answer: D

Explanation: An ecological niche refers to the specific role an organism plays within its environment, including its interactions with other species and resources. Understanding the niche of a species is crucial in agricultural practices for optimizing resource use and minimizing competition.

QUESTION 20

Answer: D

Explanation: Loam soil, a balanced mixture of sand, silt, and clay, provides optimal conditions for plant growth due to its ability to retain adequate moisture while also allowing for proper drainage, offering a suitable balance of nutrients and aeration for plants.

QUESTION 21

Answer: B

Explanation: Informational interviews provide an opportunity to connect with professionals in the field, learn about their experiences, and build valuable relationships. While it may enhance networking, it does not guarantee a job offer.

QUESTION 22

Answer: C

Explanation: Straw bales offer superior thermal insulation properties due to their high R-value, making them an eco-friendly and efficient material for constructing agricultural structures that require insulation against heat and cold.

QUESTION 23

Answer: A

Explanation: Contour plowing involves plowing across the slope following its contour lines, reducing water runoff and soil erosion by slowing down the flow of water, which allows it to infiltrate the soil more effectively, thereby minimizing erosion on steep slopes.

QUESTION 24

Answer: A

Explanation: Direct-to-consumer sales models may limit control over pricing, as consumers often expect competitive prices. In contrast, wholesale markets may offer bulk sales and pricing negotiations, providing more control over profit margins.

QUESTION 25

Answer: C

Explanation: Rapid growth during the early stages can lead to imbalances in nutrient utilization, potentially causing metabolic disorders. While it may result in larger animals, there is an increased risk of health issues associated with rapid growth, such as metabolic imbalances.

QUESTION 26

Answer: B

Explanation: Wind power is dependent on the availability of consistent and predictable wind patterns. In areas with unpredictable or insufficient wind, relying solely on wind power may result in inconsistent energy production, impacting the reliability of the energy source.

QUESTION 27

Answer: B

Explanation: Layering involves encouraging the development of roots on a stem while it is still attached to the parent plant. Once roots are established, the new plant can be separated from the parent, facilitating propagation.

QUESTION 28

Answer: A

Explanation: Consumers conducting extensive online research may have preconceived notions about products, making it challenging for sales professionals to establish personal connections. Building trust and addressing specific needs become crucial in overcoming this challenge.

QUESTION 29

Answer: D

Explanation: Monoculture systems, while efficient for specific crops, can increase the risk of pest outbreaks and disease epidemics. A lack of crop diversity creates favorable conditions for pests and diseases to thrive.

QUESTION 30

Answer: B

Explanation: Soil structure influences the arrangement of soil particles and pore spaces. Good soil structure allows for better root penetration and water movement, which are critical factors for supporting healthy crop growth.

QUESTION 31

Answer: D

Explanation: Well-draining growing media help prevent waterlogged conditions, reducing the risk of root diseases. Adequate drainage is essential for maintaining optimal plant health in greenhouse and nursery settings.

QUESTION 32

Answer: B

Explanation: Regular cleaning and sanitizing of floral design tools are essential to prevent the spread of diseases between plants. Clean tools also maintain their sharpness and functionality over time.

QUESTION 33

Answer: B

Explanation: The EPA is the primary government agency responsible for addressing environmental issues and regulations, including those related to agriculture and environmental protection.

QUESTION 34

Answer: C

Explanation: Orthographic projection is essential in shop fabrication for creating detailed and accurate representations of objects, ensuring precise measurements and layout. This technique helps in conveying a three-dimensional object onto a two-dimensional plane for fabrication purposes.

QUESTION 35

Answer: B

Explanation: A raised bed planting layout facilitates better drainage, preventing waterlogging and promoting optimal soil aeration. While raised beds can offer some weed control benefits, their primary function is to enhance drainage and create favorable conditions for root development.

QUESTION 36

Answer: C

Explanation: The growing stage is marked by rapid muscle and bone development in livestock. Adequate and balanced nutrition during this stage is crucial for ensuring proper growth and long-term health.

QUESTION 37

Answer: D

Explanation: Customer relations can be influenced by the flexibility and responsiveness of customer service. Being adaptable to customer needs contributes to positive relationships and repeat business.

QUESTION 38

Answer: A

Explanation: Isometric projection is suitable for conveying a three-dimensional understanding of a structure in a single view. It presents a distorted but proportional representation of the object, making it easier to visualize and understand. Orthographic projection, on the other hand, uses multiple views to represent an object more accurately in two dimensions.

QUESTION 39

Answer: A

Explanation: No-till farming involves minimal soil disturbance, leaving the crop residues on the surface. This method helps improve soil structure, reduce erosion, and conserve moisture. In contrast, moldboard plowing and disk harrowing involve more extensive soil disturbance. Subsoiling addresses deeper soil layers but is not focused on surface residue conservation.

QUESTION 40

Answer: D

Explanation: Environmental degradation, such as soil erosion or pollution, can lead to decreased productivity in agriculture. It may result in reduced yields and economic challenges for farmers.

QUESTION 41

Answer: D

Explanation: Oxyacetylene cutting generates heat, and rapid cooling with water after the cut can help prevent oxidation of the cut edge, preserving the metal's integrity.

QUESTION 42

Answer: C

Explanation: Sulfur is a crucial component of amino acids and vitamins involved in the photosynthesis process. While other nutrients may contribute to root development or flowering, sulfur's primary role lies in supporting the synthesis of compounds essential for photosynthetic activities in plants.

QUESTION 43

Answer: C

Explanation: Opting for recycled composite materials supports sustainability by reducing the demand for new resources. Pressure-treated wood, concrete, and vinyl may have environmental implications, but recycled composites contribute to recycling efforts and promote eco-friendly landscaping practices.

QUESTION 44

Answer: C

Explanation: While microbial fermentation does occur in the stomach and the large intestine, the cecum is a crucial site for microbial activity in livestock digestion. It serves as a fermentation chamber, aiding in the breakdown of fibrous plant materials.

QUESTION 45

Answer: B

Explanation: Contour plowing involves cultivating along the contour lines of the land, creating ridges that slow down water runoff and promote water absorption. This practice helps prevent soil erosion by reducing the velocity of water flow and enhancing soil stability.

QUESTION 46

Answer: A

Explanation: Siting agricultural structures near water sources is crucial for efficient irrigation and other water-related activities. While accessibility, aesthetics, and elevation may be considerations, proximity to water sources directly impacts the functionality and success of agricultural operations.

QUESTION 47

Answer: B

Explanation: Alfalfa is a legume crop extensively cultivated in California for its use as a forage crop and its ability to fix nitrogen in the soil, making it valuable for crop rotation and soil improvement.

QUESTION 48

Answer: C

Explanation: Heterosis, also known as hybrid vigor, is the improved or superior performance of hybrid offspring compared to their parents. This phenomenon is often observed in traits such as growth rate, fertility, and disease resistance and is a key consideration in animal breeding.

QUESTION 49

Answer: D

Explanation: Entrepreneurship in agriculture involves recognizing and seizing opportunities for innovation and growth. While risk aversion, adaptability, and resource allocation are important, the ability to identify and capitalize on opportunities is a key aspect of successful entrepreneurship.

QUESTION 50

Answer: B

Explanation: Contour plowing helps control erosion on hilly terrain by creating ridges along the contour lines, reducing water runoff and promoting water absorption. This method is particularly effective in minimizing soil erosion and maintaining stability on sloped land.

QUESTION 51

Answer: B

Explanation: Oranges, commonly harvested by hand, require delicate handling to prevent bruising or damaging the fruit. Mechanical harvesting might cause damage to the fruits, affecting their quality.

QUESTION 52

Answer: C

Explanation: Plants with deep taproot systems can access water from deeper soil layers, making them better adapted to survive drought conditions compared to plants relying on shallow root systems.

QUESTION 53

Answer: C

Explanation: Swine farrowing houses commonly use slatted floors to separate piglets from waste, facilitating waste removal and cleanliness maintenance while preventing piglets from direct contact with manure.

QUESTION 54

Answer: D

Explanation: Real estate mortgage loans typically require collateral in the form of land or property and are commonly used for long-term investments, such as purchasing land or large equipment.

QUESTION 55

Answer: C

Explanation: Disbudding or dehorning cattle is performed to minimize the risk of injuries to other animals within the herd. Horned cattle may cause harm through aggressive behavior or unintentional injuries to other animals.

QUESTION 56

Answer: D

Explanation: Prescribed burns require careful consideration and adherence to specific weather conditions to control fire behavior and prevent unintentional wildfire outbreaks, emphasizing safety and minimizing ecological impacts.

QUESTION 57

Answer: D

Explanation: Smaller handheld power equipment like chainsaws or string trimmers typically use two-stroke cycle engines due to their simplicity, lightweight, and ease of maintenance.

QUESTION 58

Answer: A

Explanation: High humidity levels can intensify heat stress in poultry by reducing their ability to dissipate heat through evaporative cooling, thus affecting their thermal comfort and health, particularly in hot climates.

QUESTION 59

Answer: B

Explanation: Cash flow analysis in agricultural business helps assess the availability of funds to meet short-term obligations, thereby evaluating the liquidity position of the business.

QUESTION 60

Answer: B

Explanation: A flail mower is designed to cut tall grass and weeds effectively while evenly distributing the cut material, making it suitable for maintaining pastures without causing excessive damage to the pasture surface.

QUESTION 61

Answer: B

Explanation: Variable Rate Technology (VRT) allows farmers to customize inputs such as fertilizers, pesticides, and water based on the specific needs of different areas within a field. This precision approach enhances resource efficiency and reduces environmental impact.

QUESTION 62

Answer: D

Explanation: Post-mortem examination involves inspecting the internal organs and tissues of slaughtered animals to detect diseases that may not be apparent externally. This process is crucial for ensuring food safety and preventing the consumption of contaminated meat.

QUESTION 63

Answer: B

Explanation: Transpiration is the process by which water is evaporated from the surface of plant leaves and stems. Its primary function is to facilitate the upward movement of water from the roots to the leaves, helping in the transportation of nutrients and maintaining plant hydration.

QUESTION 64

Answer: B

Explanation: Baby's breath is often used as a filler in floral arrangements due to its delicate, small flowers. Its airy and dainty appearance complements larger blooms and adds a light and ethereal quality to the overall design.

QUESTION 65

Answer: A

Explanation: In a Mediterranean climate characterized by hot, dry summers, drought-resistant plants with deep root systems are adapted to access water deep in the soil. This adaptation helps them survive and thrive in arid conditions.

QUESTION 66

Answer: B

Explanation: While genetically engineered crops may offer benefits such as increased resistance to pests, the potential impact on biodiversity should be a concern. The cultivation of large monocultures with genetically similar crops may reduce overall biodiversity, making the agricultural system more susceptible to pests and diseases.

QUESTION 67

Answer: C

Explanation: Rotational grazing helps reduce the risk of external parasites in cattle by preventing overgrazing in specific areas. By moving cattle to different pastures, the life cycle of parasites is disrupted, limiting their prevalence and impact on the health of the herd.

QUESTION 68

Answer: C

Explanation: Tariffs on imported agricultural products can lead to higher prices for domestic products by reducing competition from cheaper imports. This protectionist measure aims to support and incentivize local agricultural producers.

QUESTION 69

Answer: B

Explanation: When using a chainsaw, wearing eye protection is crucial to prevent injuries from flying wood chips, debris, or dust. It helps safeguard the eyes against potential hazards, ensuring the safety of the operator.

QUESTION 70

Answer: C

Explanation: Driving straight up and down the slope reduces the risk of tractor rollovers. Operating the tractor at an angle increases the likelihood of tipping, especially on slopes. This precaution promotes safe tractor operation on uneven terrain.

QUESTION 71

Answer: C

Explanation: Safety shields on equipment are designed to protect operators from moving parts. Operating the hay baler without safety shields increases the risk of accidents. Replacing the missing shields with manufacturer-approved parts ensures proper fit and function, maintaining the safety of the equipment.

QUESTION 72

Answer: B

Explanation: Black Walnut is a hardwood species native to Ohio and is prized for its high-quality wood, particularly in furniture production. Its fine-grained, dark wood is highly valued in the woodworking industry.

QUESTION 73

Answer: D

Explanation: The characteristics described—lobed leaves, milky sap, and a hollow stem—are indicative of herbaceous perennials. These plants live for more than two years, typically regrowing each spring.

QUESTION 74

Answer: B

Explanation: To showcase the diversity of agricultural products, Jessica should design interactive stations that cover various aspects of crop and livestock production. This approach provides students with a hands-on experience and a comprehensive understanding of the range of products in agriculture.

QUESTION 75

Answer: B

Explanation: Crop consultants specialize in advising farmers on crop management, utilizing precision technology like drones, sensors, and data analytics to optimize yield, manage resources efficiently, and make informed decisions in agriculture.

QUESTION 76

Answer: A

Explanation: A multimeter is a versatile tool used not only for measuring electrical properties but also for ensuring proper grounding in agricultural structures by testing the soil's electrical conductivity, which is crucial for safety and effective electrical systems in farms.

QUESTION 77

Answer: C

Explanation: High-energy diets may lead to rapid weight gain, but they can also increase the risk of metabolic disorders, such as obesity and related health issues. It's crucial to balance nutrient intake to promote healthy growth and development in animals.

QUESTION 78

Answer: B

Explanation: Shade cloth is used to regulate the amount of sunlight reaching plants, preventing excessive heat buildup. This is crucial for controlling temperature and creating optimal growing conditions in greenhouse and nursery environments.

QUESTION 79

Answer: B

Explanation: Shelterwood harvesting involves the removal of mature trees in a series of cuts, leaving some trees to provide shade and protection for natural regeneration of new seedlings. This method promotes the establishment of a new forest generation.

QUESTION 80

Answer: C

Explanation: The necessity of removing seedlings before transplanting depends on factors like the type of seedling and its root structure. While some plants may benefit from root disturbance, others may not. It's important to consider the specific needs of each plant.

QUESTION 81

Answer: B

Explanation: For effective market segmentation, Ryan should recommend dividing the market based on factors such as age, income, and lifestyle. This allows the team to tailor their marketing efforts to specific segments, increasing the likelihood of reaching and appealing to their target audience.

QUESTION 82

Answer: B

Explanation: Precision farming, also known as precision agriculture, involves using technology such as GPS, sensors, and data analytics to make more precise and efficient decisions about crop and livestock management. It aims to optimize resource use and increase productivity.

QUESTION 83

Answer: B

Explanation: Proper ventilation is essential to control humidity, remove airborne pathogens, and maintain a healthy environment. This reduces the risk of infectious diseases spreading within the animal housing facility.

QUESTION 84

Answer: A

Explanation: Low soil pH can impede nutrient uptake by plants, even if the nutrients are present in the soil. Certain essential nutrients become less available to plants in acidic soils, leading to stunted growth and nutrient deficiencies. Proper soil pH management is crucial for optimizing nutrient uptake by plants.

QUESTION 85

Answer: A

Explanation: The A horizon, commonly known as topsoil, typically contains the highest concentration of organic matter, humus, and various soil organisms, making it the most fertile layer for plant growth.

QUESTION 86

Answer: B

Explanation: While both cooperatives and partnerships involve shared ownership, a key characteristic of a cooperative is that each member, regardless of their investment or contribution, typically has an equal say in decision-making. In contrast, partnerships may distribute decision-making power based on the agreed-upon terms.

QUESTION 87

Answer: B

Explanation: Biomass for energy may lead to increased demand for land, potentially competing with the cultivation of food crops. Balancing the use of biomass for energy while ensuring food security and sustainable land management is a complex challenge.

QUESTION 88

Answer: B

Explanation: Texture in floral design involves the surface characteristics of flowers and foliage. Incorporating a variety of textures adds depth and richness to the arrangement.

QUESTION 89

Answer: B

Explanation: A needs assessment involves systematically collecting and analyzing information to determine the needs and priorities of the community and stakeholders. This information is then used to guide decision-making and program development.

QUESTION 90

Answer: A

Explanation: Yellowing of leaves, particularly in older leaves, is a common symptom of nitrogen deficiency in plants. Nitrogen is crucial for the synthesis of chlorophyll, and its deficiency can result in reduced photosynthesis and yellowing of foliage.

QUESTION 91

Answer: B

Explanation: Efficient packaging methods enhance the value-added aspect of agricultural products by ensuring better shelf life, reduced spoilage, improved transportation, and enhanced visual appeal, which can positively influence consumer perception and demand.

QUESTION 92

Answer: B

Explanation: While aptitude tests measure inherent abilities, they may not account for personal interests and values, which are crucial factors in determining career satisfaction and success. It's essential to consider a holistic approach to self-assessment.

QUESTION 93

Answer: A

Explanation: Solar power relies on sunlight, and its production can be inconsistent due to weather conditions like clouds or darkness. This intermittency can pose a challenge for maintaining a steady and reliable energy supply for farm operations.

QUESTION 94

Answer: D

Explanation: Including flowers with different heights and shapes adds interest and dimension to a floral arrangement. This diversity contributes to the overall aesthetic appeal of the design.

QUESTION 95

Answer: B

Explanation: When integrating business concepts into food science, understanding the elasticity of demand is crucial for pricing strategies. If the demand for the new product is elastic, cost-plus pricing may be suitable. If it is inelastic, penetration pricing may attract more customers.

QUESTION 96

Answer: C

Explanation: Effective communication involves fostering open dialogue, which helps address concerns and build a positive learning environment. Clear communication contributes to a supportive and collaborative atmosphere in agriscience education.

QUESTION 97

Answer: C

Explanation: Implementing proper waste management and odor control measures is essential in minimizing environmental damage from hog farms. This includes using effective waste disposal methods and technologies to mitigate odors, promoting a healthier environment for both animals and nearby communities.

QUESTION 98

Answer: B

Explanation: Organic fertilizers, while beneficial, may provide nutrients unevenly. This can lead to localized nutrient-rich areas, impacting crop growth consistency. In contrast, synthetic fertilizers often offer a more uniform distribution of nutrients.

QUESTION 99

Answer: D

Explanation: Taking a cutting involves removing a portion of a plant, usually a stem with buds, and allowing it to develop roots. This method is commonly used for the propagation of various plants, allowing the production of genetically identical individuals.

QUESTION 100

Answer: C

Explanation: Monoculture, the cultivation of a single crop over large areas, can lead to increased runoff of fertilizers and pesticides into nearby water bodies, causing environmental concerns such as water pollution.

QUESTION 101

Answer: B

Explanation: Crop rotation is a practice where different crops are planted in a specific order to disrupt the life cycles of pests and diseases, reducing their buildup in the soil. This method is a key component of integrated pest management and helps maintain soil fertility.

QUESTION 102

Answer: B

Explanation: The Leave No Trace principle advocates for minimizing human impact on natural areas during recreational activities. This includes planning and designing trails and facilities to prevent soil erosion, disturbance to wildlife habitats, and other negative effects on the ecosystem.

QUESTION 103

Answer: D

Explanation: The plasma membrane, also known as the cell membrane, is responsible for regulating the movement of substances in and out of the cell. It plays a crucial role in maintaining cell turgor pressure by controlling the passage of water and solutes.

QUESTION 104

Answer: C

Explanation: Covering manure storage helps reduce odors and limits the entry of rainwater, preventing the proliferation of pathogens and reducing environmental pollution.

QUESTION 105

Answer: C

Explanation: A grain combine is specifically designed to efficiently harvest mature crops like wheat. It cuts, threshes, and separates grain from the chaff in a single operation, minimizing grain loss and ensuring a clean and efficient harvest.

QUESTION 106

Answer: C

Explanation: Lungworms can cause respiratory issues in pigs, leading to symptoms such as coughing and labored breathing. Understanding the specific symptoms associated with different internal parasites is crucial for accurate diagnosis and effective treatment in swine production.

QUESTION 107

Answer: A

Explanation: Unity in landscape design refers to the cohesive arrangement of elements to create a harmonious and aesthetically pleasing overall design. While scale, emphasis, and proportion are also important principles, unity is particularly focused on the overall visual cohesion of the landscape.

QUESTION 108

Answer: C

Explanation: The executive summary provides a concise overview of the business plan, including the company's goals, mission, and strategies. While financial projections, marketing plans, and operations plans are critical components of a business plan, the executive summary serves as a snapshot of the entire document.

QUESTION 109

Answer: D

Explanation: Leafy greens are susceptible to pathogen contamination post-harvest. Proper handling, including hygiene practices and maintaining controlled storage conditions, is crucial to prevent foodborne illness.

QUESTION 110

Answer: C

Explanation: An increment borer is a specialized tool used in forestry to extract core samples from trees, allowing foresters to determine tree age, growth rates, and estimate the volume and value of standing timber.

QUESTION 111

Answer: B

Explanation: Temperature strongly influences the rate of plant respiration. Warmer temperatures generally lead to higher respiration rates, as enzymatic processes are more efficient. Light intensity, soil moisture, and atmospheric pressure can influence plant metabolism but are not as directly linked to respiration rates as temperature.

QUESTION 112

Answer: C

Explanation: Sustainable agriculture involves using resources in a way that meets current needs without depleting them for future generations. This approach considers environmental, economic, and social factors to ensure long-term viability and resilience in agriculture.

QUESTION 113

Answer: C

Explanation: Biological pest control involves the use of natural enemies to manage pests. In this case, releasing ladybugs to feed on aphids is an example of a biological approach to pest control. Synthetic pesticides and herbicides are chemical control methods.

QUESTION 114

Answer: C

Explanation: Habitat manipulation involves actively altering the environment to provide better conditions for targeted wildlife species. This can include activities such as planting specific vegetation or creating structures to enhance nesting sites and food sources.

QUESTION 115

Answer: B

Explanation: Stems in vascular plants often serve as storage organs, containing specialized tissues for storing carbohydrates, water, and other vital nutrients necessary for the plant's growth and development.

QUESTION 116

Answer: C

Explanation: Biosecurity protocols are designed to minimize the risk of disease introduction and spread within animal production systems, ultimately reducing the transmission of diseases among animals.

QUESTION

Answer: C

Explanation: While factors like pruning, soil pH, and phototropism can significantly affect plant growth, atmospheric pressure is not a direct factor influencing plant growth. Plants are more influenced by factors such as light, water, nutrients, and temperature.

QUESTION 117

Answer: D

Explanation: The concept of niche refers to the role or function of a species within an ecosystem. In a desert with specialized cactus species, each cactus has a specific role in the ecosystem, adapting to the arid conditions and providing unique benefits to the community.

QUESTION 118

Answer: D

Explanation: Planting design involves careful consideration of plant characteristics, ensuring that selected plants complement each other in terms of size, form, and overall aesthetics. While xeriscaping emphasizes water-efficient landscaping, planting design is specifically focused on the arrangement and compatibility of plants.

QUESTION 119

Answer: A

Explanation: Organizing in business management encompasses structuring the organization, allocating resources efficiently, and defining roles and responsibilities. It focuses on creating a framework that facilitates the achievement of organizational goals. Planning involves setting objectives, directing involves guiding activities, and controlling involves monitoring and adjusting processes.

QUESTION120

Answer: D

Explanation: Pesticide residues on crops can pose health risks to consumers if not managed properly. Adherence to proper application procedures and adherence to withdrawal periods is crucial to minimize residues in harvested crops.

QUESTION 121

Answer: C

Explanation: A clinometer is a tool used to measure angles and is particularly useful in forest management for estimating tree height or slope steepness, aiding in various forest planning and management activities.

QUESTION 122

Answer: B

Explanation: Micropropagation involves the asexual reproduction of plants, which can lead to a reduction in genetic diversity. Considering the endangered status of the plant species, the farmer should be cautious about potential loss of genetic variability, as this could impact the species' ability to adapt to environmental changes and threats.

QUESTION 123

Answer: B

Explanation: Inelastic demand means that the quantity demanded does not change significantly in response to changes in price. In this scenario, consumers are less sensitive to price fluctuations, indicating that the crop is a necessity or has few substitutes.

QUESTION 124

Answer: A

Explanation: The corpus luteum is a temporary endocrine structure formed after ovulation in the ovary. Its primary function is to produce hormones, particularly progesterone, which is essential for maintaining pregnancy and preparing the uterus for embryo implantation.

QUESTION 125

Answer: B

Explanation: A GFCI is designed to detect imbalances in electrical currents, particularly those caused by ground faults, and quickly interrupt power to prevent electrical shocks. While voltage regulation, current control, and power distribution are important aspects of electrical systems, the GFCI specifically addresses safety concerns.

QUESTION 126

Answer: D

Explanation: A sod cutter is a landscaping tool used to cut and remove strips of existing sod, making it easier to install new sod or turfgrass in landscaping projects.

QUESTION 127

Answer: C

Explanation: Market forecasting assists agricultural businesses in predicting market demand, enabling them to plan and adjust production quantities to meet market needs efficiently.

QUESTION 128

Answer: D

Explanation: Snapdragons have tall, spiky stems and a distinctive shape, making them ideal for creating lines and adding movement to floral arrangements. Their vertical structure can contribute to the overall design's dynamic and directional qualities.

QUESTION 129

Answer: B

Explanation: Artificial insemination allows for the controlled introduction of genetic material into the breeding population, leading to increased reproductive efficiency. It facilitates the use of superior sires without the need for direct physical contact between animals, improving genetic progress.

QUESTION 130

Answer: D

Explanation: Subsurface drip irrigation delivers water directly to the root zone of plants, minimizing evaporation and reducing water wastage. This method provides precise control over water application, promoting water use efficiency and minimizing the potential for soil erosion associated with other irrigation techniques.

QUESTION 131

Answer: D

Explanation: Anthelmintics are medications used to control internal parasites (worms) in livestock. They help prevent worm infestations that can negatively impact animal health and productivity.

QUESTION 132

Answer: C

Explanation: Grain combines often feature a PTO shaft that transfers power from the engine to auxiliary equipment like balers, mowers, or augers for various agricultural operations.

QUESTION 133

Answer: B

Explanation: Internal parasites, even without obvious external symptoms, can significantly impact the health and productivity of poultry. Regular monitoring and appropriate deworming measures are crucial for preventing and addressing internal parasite infestations.

QUESTION 134

Answer: B

Explanation: The variability in height and color observed in the crops is likely due to hybrid vigor, also known as heterosis. Hybrid vigor results from the increased performance and quality of plants when two different varieties are crossed, leading to improved traits.

QUESTION 135

Answer: B

Explanation: Price skimming involves introducing a new product at a high initial price to capitalize on early adopters or those willing to pay a premium, and then gradually reducing the price to attract broader customer segments over time.

QUESTION 136

Answer: C

Explanation: Autobiographies can offer personal and detailed accounts of challenges faced in specific careers. While other sources may provide valuable information, an autobiography is more likely to provide a firsthand perspective on day-to-day challenges.

QUESTION 137

Answer: C

Explanation: Conservation tillage practices, by reducing soil disturbance, can improve water infiltration and retention. This helps in preventing soil erosion and conserving water resources, contributing to sustainable soil and water management.

QUESTION 138

Answer: C

Explanation: Balance in floral design involves distributing visual weight evenly to create a sense of equilibrium. Achieving balance is crucial for creating aesthetically pleasing and harmonious arrangements.

QUESTION 139

Answer: D

Explanation: Ensuring compatibility between the welding process and the type of metal being worked on is crucial for producing a strong and durable weld. Different metals may require specific welding techniques to achieve optimal results.

QUESTION 140

Answer: B

Explanation: Sorting and grading fruits based on quality and ripeness are essential practices to minimize post-harvest losses. This ensures that only high-quality, marketable fruits are selected, reducing waste and improving overall product value.

QUESTION 141

Answer: C

Explanation: Unlike a sole proprietorship where income is reported on the owner's personal tax return, an LLC provides pass-through taxation, meaning profits and losses are passed through to the owners and reported on their individual tax returns. This allows for flexibility in tax management.

QUESTION 142

Answer: B

Explanation: Clay particles have a high surface area and a negative charge, allowing them to attract and retain water as well as nutrients. This characteristic makes clay an essential component for soil fertility and supporting plant growth.

QUESTION 143

Answer: D

Explanation: Achieving optimal weight gain is a crucial factor influencing the nutritional requirements of livestock. Factors such as species, age, pregnancy, and lactation affect the necessary nutrient intake to support healthy growth and development.

QUESTION 144

Answer: C

Explanation: Aluminum and carbon steel have different melting points and thermal conductivity. To ensure successful welding, it is essential to use a welding process suitable for each metal type. Aluminum welding often requires a specialized process, such as TIG welding, due to its unique properties.

QUESTION 145

Answer: D

Explanation: Diversified cropping involves rotating crops with different growth requirements, disrupting pest and disease cycles. This strategy helps improve soil health and reduce the need for chemical inputs, promoting sustainable crop production.

QUESTION 146

Answer: B

Explanation: When integrating agricultural knowledge into business decision-making, understanding plant physiology is crucial. It influences the growth, development, and yield of crops, which in turn impacts the financial outcomes of the farming practices. This knowledge helps in making informed decisions related to crop production.

QUESTION 147

Answer: C

Explanation: Effective leadership in agriscience education involves providing constructive feedback and support for the professional development of teachers. This approach enhances the overall effectiveness of the educational team.

QUESTION 148

Answer: C

Explanation: The safest course of action is for the student to investigate and troubleshoot the issue before attempting to start the tractor again. Bypassing safety features or repeatedly attempting to start without identifying and addressing the problem can lead to unsafe conditions.

QUESTION 149

Answer: A

Explanation: Conventional agriculture often involves extensive tillage, heavy chemical inputs, and monoculture practices, which can lead to increased soil erosion and nutrient runoff, negatively impacting the environment compared to other production systems.

QUESTION 150

Answer: C

Explanation: Relying on a single insecticide or rotating those with similar modes of action can contribute to insecticide resistance. Introducing natural predators as part of an integrated approach helps control pest populations without promoting resistance.

QUESTION 151

Answer: B

Explanation: In the negotiation process for an agriculture-related position, allowing the employer to propose a salary range first provides valuable information. It helps in understanding the employer's expectations and allows for a more informed and effective negotiation.

QUESTION 152

Answer: C

Explanation: An Entrepreneurship SAE involves students in planning, implementing, and operating an agricultural business. This could include activities such as raising and selling livestock, producing and selling agricultural products, or providing services related to agriculture.

QUESTION 153

Answer: C

Explanation: Biosecurity encompasses a range of measures, including vaccination and maintaining optimal environmental conditions, to prevent the introduction and spread of diseases within an animal population.

QUESTION 154

Answer: B

Explanation: While textbooks and traditional media may become outdated, online forums with industry professionals provide real-time insights and discussions on current trends, challenges, and opportunities in the agriculture job market.

QUESTION 155

Answer: B

Explanation: Burning agricultural plastic waste can release toxic fumes and air pollutants, posing environmental and health risks. It's essential to explore alternative waste management practices to mitigate these negative impacts on air quality.

QUESTION 156

Answer: A

Explanation: Supervised agricultural experiences (SAEs) are a crucial component of the total agricultural program model, designed to enhance students' learning by engaging them in practical, hands-on activities related to agriculture.

QUESTION 157

Answer: B

Explanation: The deep-litter system involves the accumulation of bedding material in the poultry coop. This system helps reduce ammonia levels by promoting microbial decomposition of manure, providing a healthier environment for the birds and minimizing respiratory issues.

QUESTION 158

Answer: C

Explanation: The availability of oxygen in the root zone is essential for the process of respiration in plants. During respiration, plants use oxygen to break down organic compounds and release energy. Insufficient oxygen in the root zone can negatively impact this vital physiological process.

QUESTION 159

Answer: B

Explanation: Monoculture practices involve cultivating a single crop repeatedly in the same area, leading to increased vulnerability to pests and diseases, soil degradation, and reduced biodiversity. It contradicts sustainability principles by intensifying the risks associated with agriculture.

QUESTION 160

Answer: B

Explanation: Drip irrigation involves a higher initial investment due to its intricate system of tubes and emitters, yet it provides precise water delivery directly to plant roots, minimizing water waste and optimizing water usage in agriculture.

QUESTION 161

Answer: C

Explanation: Inbreeding, or mating closely related animals, can increase the likelihood of genetic disorders due to the expression of recessive alleles. While it may lead to desired traits, it also poses a risk of undesirable genetic conditions.

QUESTION 162

Answer: A

Explanation: The tomato is often classified as both a fruit and a vegetable due to its culinary uses. While scientifically a fruit (it develops from the ovary of a flower and contains seeds), it is commonly used in savory dishes, leading to its vegetable classification in a culinary context.

QUESTION 163

Answer: B

Explanation: The Law of Supply and Demand is fundamental in agribusiness decision-making. In this case, understanding the demand for organic tomatoes and how it may affect pricing and market conditions is crucial for making informed choices.

QUESTION 164

Answer: C

Explanation: Effective equipment maintenance involves a holistic approach. Regular lubrication ensures mechanical components function smoothly, and proper calibration of electronic components maintains accuracy and efficiency.

QUESTION 165

Answer: A

Explanation: Soilless growing media provide better control over nutrient content, allowing for precise management of plant nutrition. This is especially important in greenhouse and nursery production for optimizing plant growth.

QUESTION 166

Answer: C

Explanation: Cultural factors can significantly influence dietary preferences and habits, creating challenges for marketing agricultural products internationally. Understanding and adapting to these variations are crucial for successful international trade.

QUESTION 167

Answer: D

Explanation: Sustainable management practices aim to meet the needs of visitors while minimizing negative impacts on the environment. This involves carefully planning and implementing measures to ensure the long-term health and integrity of outdoor recreation areas, considering ecological, social, and economic factors.

QUESTION 168

Answer: B

Explanation: Foliar spraying involves applying chemicals directly onto plant foliage. It carries a higher risk of drift due to wind, potentially causing unintended contamination of nearby areas or non-target plants compared to other application methods.

QUESTION 169

Answer: B

Explanation: Economic incentives, such as government subsidies or market demand for sustainably sourced products, significantly influence forest management practices, encouraging conservation efforts and sustainable forestry practices.

QUESTION 170

Answer: B

Explanation: Photosynthesis primarily takes place in the chloroplasts of plant cells. These organelles contain chlorophyll, the pigment responsible for capturing light energy and converting it into chemical energy through the photosynthetic process.

QUESTION 171

Answer: A

Explanation: Plasma cutting is known for its precision and ability to create intricate designs. It uses a high-velocity stream of ionized gas to cut through metal with greater accuracy compared to oxyacetylene cutting. This makes plasma cutting the preferable choice for detailed metal fabrication work.

QUESTION 172

Answer: C

Explanation: Feathering is a term associated with poultry, specifically the presence of feathers. In livestock carcass evaluation, factors like marbling (intramuscular fat), muscling, and yield grade (amount of lean meat) are considered, but feathering is not relevant to non-poultry carcasses.

QUESTION 173

Answer: C

Explanation: A dado joint involves cutting a groove across the grain of one piece of wood to receive the end of another piece. This joint provides increased surface area for gluing, resulting in a strong and durable connection. While other joints serve different purposes, the dado joint specifically enhances the bond between pieces.

QUESTION 174

Answer: C

Explanation: Hedge trimmers are specialized tools equipped with reciprocating blades designed for precisely trimming and shaping shrubs and hedges in landscaping.

QUESTION 175

Answer: D

Explanation: Democratic leadership involves fostering a collaborative and participative approach, with input from team members in decision-making. While autocratic leadership is more authoritative, laissez-faire leadership allows for autonomy, and transformational leadership focuses on inspiring and motivating, democratic leadership emphasizes inclusivity and shared decision-making.

QUESTION 176

Answer: B

Explanation: Drip irrigation delivers water directly to the plant root zone, minimizing water loss due to evaporation or runoff, making it one of the most water-efficient methods for irrigating landscape plants.

QUESTION 177

Answer: C

Explanation: A faculty advisor's role in a student agricultural organization involves mentoring and assisting students in developing and achieving the organization's goals, fostering leadership and organizational skills.

QUESTION 178

Answer: D

Explanation: Automated drip irrigation systems with real-time data monitoring optimize irrigation efficiency by delivering precise amounts of water based on current environmental conditions. This technology helps conserve water resources and enhance crop yields.

QUESTION 179

Answer: D

Explanation: Implementing blockchain technology offers transparent and secure traceability throughout the supply chain, enhancing the value-added principles by ensuring authenticity, quality, and accountability from production to distribution in the agricultural sector.

QUESTION 180

Answer: C

Explanation: Modern combines often have an internal combustion engine for propulsion and power generation, and electrical systems for various functions such as monitoring, control, and precision agriculture features.

QUESTION 181

Answer: D

Explanation: The mitochondrion is the cellular organelle responsible for the synthesis of ATP through cellular respiration. While chloroplasts are involved in photosynthesis, mitochondria play a crucial role in energy production within plant cells.

QUESTION 182

Answer: B

Explanation: In agribusiness, especially when adopting new technologies, a sensitivity analysis is crucial. This involves assessing how changes in various factors, such as the effectiveness of the technology or market conditions, may impact the overall outcome. It helps in making more informed and robust decisions in the face of uncertainty.

QUESTION 183

Answer: B

Explanation: Horses are hindgut fermenters as they possess a specialized digestive system where microbial fermentation primarily occurs in the hindgut (cecum and colon), aiding in the breakdown of fibrous plant material that was not fully digested in the foregut (stomach and small intestine).

QUESTION 184

Answer: C

Explanation: Consumers focused on sustainability are likely to be influenced by environmentally friendly practices. Highlighting eco-friendly certifications and sustainable practices aligns with the values of these consumers and can contribute to the decision-making process.

QUESTION 185

Answer: C

Explanation: Clayey soil has the smallest particles among the options, providing a high surface area for nutrient retention. This soil type is known for its fertility, although it may have challenges related to drainage.

QUESTION 186

Answer: A

Explanation: Rotational grazing involves periodically moving livestock between different grazing areas. This practice enhances pasture productivity by allowing for rest and regrowth of vegetation, optimizing forage availability for the animals.

QUESTION 187

Answer: B

Explanation: Wind speed affects both transpiration and photosynthesis. Increased wind speed can enhance transpiration by promoting water vapor loss from leaves, and it can also influence the efficiency of photosynthesis by ensuring a fresh supply of carbon dioxide around the leaves.

QUESTION 188

Answer: B

Explanation: The primary purpose of a cash flow statement in agribusiness accounting is to track the inflow and outflow of money. This statement helps businesses manage their liquidity and make informed financial decisions.

QUESTION 189

Answer: C

Explanation: Transformational leaders in agriscience education inspire and motivate others, encouraging students and colleagues to achieve beyond their perceived limitations. This leadership style fosters growth and innovation.

QUESTION 190

Answer: B

Explanation: Implementing rotational grazing practices is a sustainable strategy to reduce greenhouse gas emissions in cattle production. Rotational grazing allows for better land management, enhances soil health, and reduces the environmental impact associated with conventional grazing methods.

QUESTION 191

Answer: D

Explanation: The proventriculus in avian anatomy is equivalent to the human stomach and has similar functions to the liver. It secretes digestive enzymes and gastric juices, playing a crucial role in the initial digestion of food in birds.

QUESTION 192

Answer: C

Explanation: Fungi are primary decomposers in soil, breaking down complex organic materials into simpler forms. Their activity contributes to the formation of humus, which is vital for soil structure and nutrient cycling.

QUESTION 193

Answer: B

Explanation: Overuse of nitrogen-based fertilizers can lead to lush, succulent plant growth, making crops more susceptible to certain diseases. This excessive growth may weaken the plants' natural defenses, creating favorable conditions for diseases to thrive.

QUESTION 194

Answer: A

Explanation: Sexual reproduction involves the fusion of genetic material from two parent plants, resulting in greater genetic diversity among offspring. This diversity can be advantageous for adapting to changing environmental conditions.

QUESTION 195

Answer: B

Explanation: In a resume for an agricultural career, creating a distinct section exclusively for technical skills allows the candidate to highlight their specific expertise clearly, making it easier for potential employers to identify relevant qualifications.

QUESTION 196

Answer: B

Explanation: Blockchain technology is increasingly used in agribusiness to enhance transparency and traceability throughout the supply chain. It allows for secure and transparent recording of transactions, ensuring the authenticity and origin of agricultural products.

QUESTION 197

Answer: C

Explanation: Improving air circulation and ventilation is crucial in greenhouse management to reduce humidity and prevent fungal diseases. Proper airflow helps maintain a healthy environment for plants, minimizing the risk of diseases caused by excessive moisture.

QUESTION 198

Answer: C

Explanation: The Risk-Return Tradeoff is essential when choosing between financing options. Students must weigh the potential returns against the associated risks. Traditional bank loans may offer lower returns but with lower risks, while venture capital investments may provide higher returns but come with higher risks.

QUESTION 199

Answer: B

Explanation: The central vacuole in plant cells is primarily responsible for storing nutrients, waste products, and maintaining turgor pressure. It plays a crucial role in regulating cell structure and function.

QUESTION 200

Answer: B

Explanation: Effective communication in agriscience education involves adapting language and presentation to the audience's knowledge level. This ensures that information is conveyed clearly and promotes better understanding among students.

QUESTION 201

Answer: A

Explanation: Perceived scarcity, or the idea that a product is limited in quantity, often creates a sense of urgency and can significantly influence consumer behavior. This psychological factor taps into the fear of missing out and can drive consumers to make a purchase.

QUESTION 202

Answer: C

Explanation: Poultry, such as chickens and birds, possess a brood patch—a specialized area on the abdomen where feathers are shed and blood vessels are close to the skin. This patch helps to transfer heat to the eggs during incubation, aiding in their development.

QUESTION 203

Answer: B

Explanation: While automated data collection systems offer efficiency, they may have limitations in adapting to rapidly changing environmental conditions. Human intervention is crucial for interpreting unpredictable factors and making real-time adjustments in farm management.

QUESTION 204

Answer: C

Explanation: Implementing strict hygiene and sanitation measures is crucial in preventing the spread of foodborne pathogens in crop production. This includes proper handling, cleaning, and storage practices to ensure the safety of the harvested crops.

QUESTION 205

Answer: B

Explanation: Adding insulation to the greenhouse walls and roof is a structural modification that can help regulate temperature by reducing heat loss. This improves the energy efficiency of the greenhouse and provides a more stable environment for plant growth.

QUESTION 206

Answer: D

Explanation: Nutrient cycling refers to the movement and recycling of nutrients within an ecosystem, where organic matter decomposes, releasing nutrients back into the soil, ensuring the availability of essential elements for plant growth and sustaining agricultural productivity.

QUESTION 207

Answer: A

Explanation: Hybridization involves the intentional crossbreeding of two different individuals to combine desirable traits. This method is widely used in agriculture to develop crops with improved characteristics.

QUESTION 208

Answer: D

Explanation: Bedding plants are often used to provide mass displays of color in flower beds and containers. Petunias are a popular choice for bedding plant displays due to their vibrant colors and ease of cultivation.

QUESTION 209

Answer: D

Explanation: The nutritional value of feed is primarily determined by its nutrient content, including proteins, carbohydrates, fats, vitamins, and minerals. These components play a key role in meeting the dietary needs of livestock.

QUESTION 210

Answer: B

Explanation: Weight loss is a noticeable symptom of nutrient deficiency in livestock. Agriculture teachers should be vigilant in recognizing signs of inadequate nutrition to address and correct deficiencies in the animals' diet.

QUESTION 211

Answer: A

Explanation: Sustainable agriculture promotes crop rotation as a practice to maintain soil health, prevent pest build-up, and reduce the need for chemical inputs. This approach contributes to minimizing environmental degradation.

QUESTION 212

Answer: D

Explanation: Identifying target markets involves tailoring agricultural products to meet the specific needs and preferences of consumers. This customization enhances the likelihood of success in marketing.

QUESTION 213

Answer: C

Explanation: Governmental factors can significantly affect international trade by introducing trade barriers, regulations, and policies that impact the movement of agricultural products across borders.

QUESTION 214

Answer: D

Explanation: Wood is combustible, which makes it unsuitable for tasks involving high temperatures or sparks in metalworking. In such cases, non-combustible materials like metals are preferred to ensure safety in the fabrication process.

QUESTION 215

Answer: B

Explanation: Phosphorus plays a crucial role in promoting flower and fruit development, as well as overall root growth and energy transfer within the plant. While nitrogen is essential for leafy green growth, and potassium contributes to overall plant health, phosphorus specifically targets reproductive processes.

QUESTION 216

Answer: D

Explanation: Soil fertility assessments commonly include pH level, organic matter content, and electrical conductivity. However, atmospheric pressure is not a standard parameter for soil fertility testing.

QUESTION 217

Answer: B

Explanation: Granular fertilizers release nutrients slowly as the granules break down over time. This controlled release is beneficial for long-term plant nutrition. In contrast, liquid fertilizers offer a quicker nutrient uptake, making granular fertilizers more suitable for sustained nutrient supply.

QUESTION 218

Answer: B

Explanation: Genotype refers to the genetic makeup of an individual. When evaluating livestock for breeding, understanding the genotype is crucial as it determines the animal's ability to pass on desirable traits to its offspring. Phenotype, heterosis (hybrid vigor), and heritability are important concepts in breeding but focus on different aspects of genetic expression and inheritance.

QUESTION 219

Answer: B

Explanation: A P-trap is a U-shaped pipe in plumbing that retains a small amount of water to prevent the backflow of gases and sewage from the drainage system into the living or working space. While other functions like regulating pressure, facilitating drainage, and removing impurities are important in plumbing, the P-trap specifically addresses backflow prevention.

QUESTION 220

Answer: B

Explanation: Holstein cattle are renowned for their high milk production capacity, often used in dairy operations worldwide, producing milk with relatively high butterfat content.

QUESTION 221

Answer: C

Explanation: Hedging involves using futures contracts or options to protect against adverse price movements by locking in prices for future transactions, thus minimizing the impact of market volatility.

QUESTION 222

Answer: B

Explanation: Tractors frequently employ a CVT system that offers infinite gear ratios, allowing the engine to operate at its most efficient speed regardless of the tractor's ground speed, optimizing fuel consumption and engine performance.

QUESTION 223

Answer: C

Explanation: In California, where water conservation is crucial, selecting plants with high drought tolerance is essential for sustainable landscaping. While factors like growth rate and flower color may be considerations, prioritizing drought-tolerant species aligns with water-wise landscaping practices.

QUESTION 224

Answer: C

Explanation: Independent assortment during meiosis leads to the random distribution of homologous chromosomes into different gametes, contributing to genetic diversity in the offspring. This process ensures that each gamete receives a unique combination of genetic material.

QUESTION 225

Answer: B

Explanation: A ridge vent allows hot air and moisture to escape from the building, promoting natural ventilation. This design feature helps maintain optimal air quality by preventing the buildup of stale air and moisture within the livestock building.

QUESTION 226

Answer: C

Explanation: Foot-and-Mouth Disease (FMD) in livestock is characterized by the appearance of vesicular lesions or blisters on the mouth, feet, and sometimes on the udder. These blisters can cause lameness and difficulty in feeding.

QUESTION 227

Answer: B

Explanation: The drive shaft is a key component in the power transmission system of agricultural machinery, transferring power from the engine to various machinery components such as the gearbox, wheels, or implements.

QUESTION 228

Answer: D

Explanation: Maintaining a proper energy balance is crucial during gestation to ensure the nutritional needs of the developing fetus. While protein, calcium, and other nutrients are important, achieving the right balance of energy is key for supporting healthy fetal growth and development.

QUESTION 229

Answer: D

Explanation: Monocots typically exhibit parallel venation in their leaves, while dicots have netted or branching venation. This distinction in leaf venation is a key feature used in the classification of these plant groups.

QUESTION 230

Answer: B

Explanation: A rotary tiller is ideal for preparing seedbeds in rugged or uneven terrain. It efficiently breaks up soil clods, incorporates organic matter, and creates a smooth, fine seedbed, making it suitable for pre-planting cultivation.

QUESTION 231

Answer: C

Explanation: Agricultural cooperatives allow farmers to pool resources, collectively negotiate better prices for inputs and outputs, and gain stronger bargaining power in the marketplace. This can lead to improved economic outcomes for individual farmers.

QUESTION 232

Answer: B

Explanation: GIS and GPS technologies enable farmers to precisely map and target areas within a field, allowing for the uniform application of inputs such as pesticides. This targeted approach can reduce pesticide use, minimize environmental impact, and optimize crop protection.

QUESTION 233

Answer: D

Explanation: To maintain stability and prevent accidents, it's crucial to distribute the weight evenly across a trailer. This ensures proper balance and reduces the risk of tipping or swaying during transportation.

QUESTION 234

Answer: D

Explanation: Clean water and sanitation are critical for preventing bacterial growth in the vase water, which can lead to premature wilting and decay of cut flowers. Proper hygiene practices, such as regularly changing water and cleaning containers, contribute to the overall health and longevity of the floral arrangement.

QUESTION 235

Answer: D

Explanation: Using the correct tool for the intended task not only ensures efficiency but also minimizes the risk of accidents. Using tools for purposes they were not designed for can lead to breakage or personal injury.

QUESTION 236

Answer: B

Explanation: The efficiency of wind power generation depends on the availability and consistency of wind. Higher and more consistent wind speeds are essential for optimal energy production from wind turbines.

QUESTION 237

Answer: C

Explanation: Burning or burying pesticide containers can lead to environmental contamination. Recycling through an approved program is the safest and most environmentally responsible way to dispose of pesticide containers, ensuring proper handling of residual chemicals.

QUESTION 238

Answer: B

Explanation: Subsidies can incentivize farmers to increase production of the subsidized crop, leading to overproduction and surplus. This surplus can result in lower market prices and potential storage challenges, impacting the overall stability of the agricultural market.

QUESTION 239

Answer: B

Explanation: Harvesting cut flowers at the right stage of development ensures better vase life. Picking flowers at the optimal time, such as when buds are just beginning to open, can lead to longer-lasting and more attractive floral displays, benefiting both the florist and the customer.

QUESTION 240

Answer: A

Explanation: Storing fuels in a well-ventilated area is critical to prevent the buildup of fumes, which can be flammable and pose a safety risk. Proper ventilation reduces the likelihood of accidents and ensures a safe working environment.

www.ingramcontent.com/pod-product-compliance
Lightning Source LLC
LaVergne TN
LVHW082357200926
848481LV00012B/3598